Conflict, Environment, and Social Complexity

Series Editor

Richard J. Chacon, Sociology & Anthropology Department, Winthrop University, Rock Hill, USA

This series aims to expand our understanding of conflict, anthropogenic environmental impacts, and the rise of complex societies from a host of interdisciplinary approaches and theoretical perspectives. The series seeks publications on the origins and nature of prehistoric conflict, warfare, and ritual violence broadly understood as well as how conflict, warfare, and ritual violence manifest themselves in modern day contexts. This series also examines how prehistoric patterns of natural resource utilization impacted ancient environments as well as how ongoing use of natural resources affects present day environments. Additionally, it analyzes the internal and external factors that gave rise to social complexity from a variety of materialist and non-materialist theoretical perspectives. The broad scope of the series encompasses prehistoric, historical, and contemporary case studies from multiple geographical regions. Lastly, the series' multidisciplinary approach welcomes archaeological, ethnographic, historical, sociological, political science, biological, conservationist, land use policy, and environmental perspectives that will produce new and valuable insights on salient topics. The benefits from these new insights will extend beyond the academic world to society as a whole.

Christopher S. Davis

Before Sunset

Ice-Age Amazonian Rock Art and Archaeoastronomy at the Younger Dryas

Springer

Christopher S. Davis
Archaeologist, Anthropology Department
McHenry County College
Chicago, IL, USA

ISSN 2730-5872 ISSN 2730-5880 (electronic)
Conflict, Environment, and Social Complexity
ISBN 978-3-031-93372-1 ISBN 978-3-031-93373-8 (eBook)
https://doi.org/10.1007/978-3-031-93373-8

This Springer imprint is published by the registered company Springer Nature Switzerland AG
The registered company address is: Gewerbestrasse 11, 6330 Cham, Switzerland

This book is dedicated to my loving wife Carmen and my wonderful and spiritually radiant children, Amani, Khalil, and Bakari. May peace and good friends accompany you through success and a promising future. "Our lives are not our own. From womb to tomb, we are bound to others. Past and present. And by each crime and every kindness, we birth our future"

– Sonmi~451

From Cloud Atlas, Mitchell, David. Random House, 2004; Screenplay by Lana Wachowski, Tom Tykwer, and Andy Wachowski, Warner Bros. Pictures, 2012

Preface

How did human ancestors survive the Pleistocene, a nearly two-million-year period of advancing and retreating mile-high glaciers and meters-tall megafauna? How did people settle into the Americas so fast and so far during the last ice age? A person with knowledge of survival skills can wander far in the span of their life, but to establish successfully sustained settlements requires a community of reproductive mates, food to thrive on, and safety in which progeny can mature. For a founding community to do so within travel distance of their place of origin is how scholars imagine settlements expand, which typically estimates to a few kilometers per generation, but how this was done in the Americas when the earliest pioneer communities were presumably an entire hemisphere away from their place of origin is perplexing.

One such place in the Americas was revealed, not in Alaska near the Bering land bridge, nor along the coast where small boats could have skirted along the shore, but in the interior of South America, near the Amazon River, over 11,000 uncalibrated years before present—roughly 13,000 calendar years ago. Therefore it is both a curiosity and also a necessity that modern societies gleam into the lifestyles of these early inhabitants of the lowland Brazilian Amazonian rainforest, because they settled a pristine environment at a critical juncture of geological climate transitions similar to those we now face from our fossil fuel ramifications. How did Paleoindians in South America successfully adapt to a new ecosystem during a new geological epoch—the Younger Dryas?

I believe ethnoastronomy traditions in the Americas may have become more sophisticated in tropical zones because people there would have relied more heavily on these observations to mark the passage of time than people living in higher latitude environments where seasonal transitions are more overt (daylight hours, foliage, temperature). Water travel is another way of life that vastly benefits from astronomy-assisted navigation. Coastal cultures may have brought this knowledge inland to the interiors of the Amazon and Orinoco River basins at an early time of human habitation.

If this theory is true, we might expect to find other modes of astronomy notations among coastal Paleoindian cultures in North and South America, as we have with much more recent pioneering cultures like in Polynesia. There should be more research investigations into coastal pacific North America and Hawaii using museum collections and archaeology reports to search for signs of incipient or terrestrial-adapted astronomy lore in their material records. Generating new data on these topics will enhance our understanding of how humans adapted their traditional knowledge to more effectively and successfully colonize new environments.

It seems like an unlikely place. Just west of an ubiquitous bend along the Amazon River in Brazil's rainforest, far from the river's Andean mountain source, and yet still far from the river's terminus in the Atlantic ocean, are several low-lying hills that were renamed *Monte Alegre* by its eighteenth-century European inhabitants in order to encourage settlers to the region. It is ironic that settlers had to be encouraged to come here, because Paleoindians inhabited the area (seemingly seasonally) thousands of years earlier, and they remained for at least several hundred years. Later groups reinhabited the land in periods separated thousands of years apart.

Even at the time of European contact, several native tribes lived along the river near the hills; they had tribal names like the *Gurupatuba* and the *Buzibizes*. However, the only name given to those natives who lived in the hills at that time were the *Serranos*, a generic Portuguese moniker that translates to "hill folk." In the early decades of European habitation, a mission was first established, and it coralled the surrounding native peoples into its confines as laborers—or slaves—the relationship is not fully confirmed. Already at this time, those *Serranos* who lived in the hills were possibly refugees, or sought refuge, and were possibly not the original inhabitants.

Indeed, archaeological evidence suggests that the valleys around the hills were inhabited four hundred to one thousand years ago, but not the caves, whose last evidence of habitation roughly dates from about four to two thousand years ago. And so the scantily written records describing the hills and its rock art were (stated or inferred by the European chroniclers) to come from the mission Indians—who themselves almost certainly surmised about the artistic subjects depicted: animals like possums, manatees, anteaters, jaguars, monkeys, and sun and moon images.

Other chronicled references about the art were wholly deduced by European naturalists like Alfred Russel Wallace and Charles Francis Hartt to be "some form of sun worship, especially in the vicinity of the rock tower." This idea must have been widely adopted by the Brazilian inhabitants of the region in the coming centuries, who renamed the artistic locations in the hills: Serra da Lua and Serra do Sol—hill of the moon and hill of the sun. Native tribal names for the vicinity, especially the hills with art in them, were also passed down through time, but only in the Tupí language, which had only become a commonly used language in the region shortly before European arrival, and therefore were probably a rough translation at best. Those names were *Erere* and *Paituna*.

Paituna is the name of the hill ridge (containing art and numerous caves) that lies closest to the Amazon River, and the name refers to the "black water lake" (though really a lagoon) that forms to the west of this hill during the rainy season. The other hill ridge, *Erere*, is higher and drier, with only three known caves. The meaning of the word is lost to most Brazilian scholars today; however, its most prominent cave,

Itatupaoca, means stone house of the (father) lord. The Tupí often utilized the word *tupa* or *tuba*, which meant lord or deity, in their names. However, this provides further clues to suggest that even these indigenous names of the region are themselves foreign, because the early missions reported nearly all of the regional tribes to have a religion that worshipped female nature deities of the sun and moon: *Jací* and *Coarasí*. The overall inference anyone should draw from the place names is that the artists are long removed in time, and perhaps culture, from the historic-period inhabitants.

So this book presents the current landscape and known history of the region first. Following this is a summary description of the rock art motifs and locations in the hills, and the archaeological evidence that dates it to the Pleistocene. The latter half of the book presents excavation data, artifacts, landscape features, and arguments for the possible significance and/or motivation for the art, and the legacy it appears to have left or shared commonly with many Native American communities through North and South America by the time of historical record. Perhaps most surprising to me was the slow realization that the earliest inhabitants engaged in substantial landscape modification at *Painel do Pilão*, which ends at the time that the Younger Dryas begins. There is even the possibility of a slight overlap of a generation or so, giving much greater weight to the reason for such activity in art. Could the art have captured the actual cause of the Younger Dryas?

Chicago, IL, USA Christopher S. Davis

Acknowledgments

I am forever appreciative of my Dissertation Director, Dr. Anna Roosevelt, who plucked me out of graduate school limbo and placed me in front of the grand designs of ancient Amazonian artists. I would also like to express my appreciation to Dr. Laura Junker who continually informed, encouraged, and assisted me to apply for grants and funding. Appreciation is extended to my other committee members Dr. Joel Palka, who remained in my corner since the beginning, and Dr. Waud Kracke who inspired me to explore my dreams as much as he assured me that such esoteric investigation was worthy of academic study. Additionally, my preliminary committee members Dr. Antonio Curet, Dr. Robert Hall, and Dr. Virginia Miller deserve my appreciation as well for helping me climb to a new base from which to start a higher ascent. I would like to further acknowledge Dr. Jack Prost for initiating my sense of investigative research and Dr. Mark Lichty for teaching me the ropes when I was a newcomer to the discipline of anthropology.

Equally valuable are all those who helped me in Brazil, beginning with Dr. Denise Schaan as my foreign advisor who signed off on all my projects and provided the opportunity for my first true publication. A true heartfelt appreciation is extended to Carmen Silvia Viana Trindade for her truly dedicated efforts in the field and at IPHAN, and in general for being my *protector*. Marcio Amaral and Leonard Grala were indispensable assistants every time I returned to the field. Senhor Tiba and Nelci Sadek deserve my deep gratitude for their transportation and guide services. Senhor Ozias and his family had become my family away from home.

Eternal lifelong gratitude is reserved for my own family, my parents Shirley Rounds and Lorenzo Davis, who instilled in all their children the importance of education. When they weren't guiding my development directly, my four older siblings Lorenzo Jr, Aaron, Julian, and Teresa stepped in by example. You all were my first teachers. To my friends, colleagues, and students too numerous to name, you all honed my thoughts with your deep philosophical discussions, arguments, and experiences.

There are so many of my family, friends, and coworkers who have been instrumental in my life that it would take another book to name them all. So, these acknowledgments are focused on those who were instrumental in this Monte Alegre research. I am deeply grateful to my dissertation director, Dr. Anna Roosevelt, for guiding me from the uncertainties of graduate school to the profound insights of ancient Amazonian artistry. My heartfelt thanks also to Dr. Laura Junker, whose continuous support and encouragement were instrumental in securing grants and funding.

I extend my sincere appreciation to my committee members: Dr. Joel Palka for his unwavering support since the beginning of my graduate education; Dr. Vincent LaMotta for exemplifying the end goal, Dr. Virginia Miller for the continuous reminder that there is more than one way to perceive the world, and the late Dr. Waud Kracke for inspiring me to pursue my dreams and validating the academic worth of my esoteric investigations. Thanks, also, to my preliminary committee members, Dr. Robert Hall and Dr. Antonio Curet, for providing a solid foundation for my academic ascent. Special thanks to Dr. Jack Prost for sparking my interest in investigative research and to Dr. Mark Lichty for broadening my scope of anthropology past the material artifacts to the people that imbued them with meaning. And, of course, thanks to Dr. John Monaghan for overseeing my graduate years as department chair.

In Brazil, I am indebted to Dr. Denise Schaan for endorsing my projects and facilitating my first publication. My profound gratitude goes to my wife Carmen, who I met during my research in Brazil, for her dedicated help in the field and at IPHAN. I am also grateful to Marcio Amaral and Leonard Grala for their indispensable assistance during my fieldwork and to Senhor Tiba and Nelci Sadek for their transportation and guidance services. Senhor Ozias and his family, Nildo, Obede, Silvia, and Dona Creusa became my family away from home, and I am deeply thankful for their hospitality.

Eternal gratitude is reserved for my family. My parents, Shirley and Lorenzo, who instilled in all their children the importance of education as demonstrated in my siblings, Lorenzo Jr., Aaron, Julian, and Teresa. You have all guided my development by example and were my first teachers. To my extended family, friends, colleagues, and students, too numerous to name, thank you for sharpening my thoughts through deep philosophical discussions, arguments, and shared experiences. Hopefully there will be future books where I can name more of you directly.

Competing Interests The author has no competing interests to declare that are relevant to the content of this manuscript.

Contents

Chapter 1
Introduction: Monte Alegre Rock Art

Approximately fourteen kilometers southwest of the modern-day city of Monte Alegre, in the Brazilian state of Pará, more than fourteen ancient rock art sites dot the caves and rock faces of three sandstone hill ridges next to the northern bank of the Amazon River (see Fig. 1.1). The hill ridges (called *Serras*) had been formed during the Paleozoic to the Tertiary geological time periods, and they form the southern arc of an ancient volcanic dome which the Brazilian government has designated as a state park [*Parque Estadual Monte Alegre*—PEMA (see Milhomens et al. 2008)]. These hills of mostly sedimentary composition, called *Serra do Ererê*, *Serra da Paituna*, and *Serra da Maxirá*, offer numerous caves, rock shelters, and open-air cliffs, some of which prehistoric artists decorated with mostly red—but also several yellow, and some red and yellow combined—pictographs usually overlaying a painted white background.

Nearly three hundred of the approximately one thousand recorded images are found at a single site, *Serra da Lua*, which stretches about one kilometer along the northwest curve of Serra do *Ererê.* Fewer than ten of these images can be classified as true *petroglyph* engravings. Instead, many are abrasions on rock surfaces, creating depressions or enlarging and smoothing natural orifices. Four of the rock art sites consist of open-air rock faces, while the remaining sites are caves or shallow rock shelters. Nearly all of the open-air sites contain both individual paintings and panels of grouped *pictographs,* while most of the caves feature only individual paintings.

Caverna da Pedra Pintada is the largest painted cave among all the rock art sites. Dr. Anna Roosevelt (1996) and her team excavated there and unearthed evidence of a late Pleistocene Paleoindian occupation period separated from a more recent Holocene occupation period by a sterile layer corresponding to about a 3000-year time span of presumed site abandonment. Only one fragment of red ochre was found in association with the Holocene layers that were dated between 7500 and 1450 years ago, but numerous ochre fragments, paint drops, and lumps of pigment were found with Paleoindian artifacts and black soil in the oldest stratigraphic

C. S. Davis, *Before Sunset*, Conflict, Environment, and Social Complexity,
https://doi.org/10.1007/978-3-031-93373-8_1

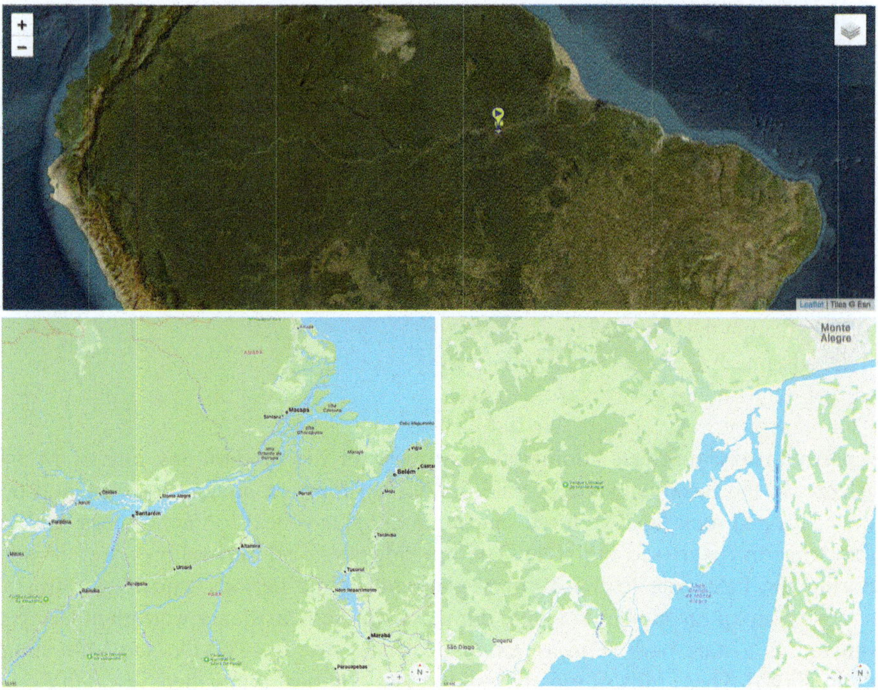

Fig. 1.1 Monte Alegre (yellow pin in top photo) along the Amazon River in the Brazilian state of Pará, which lies southwest of the Brazilian coastal state of Amapa (bottom left). The hills in the state park (PEMA) are southwest of the town of Monte Alegre (bottom right)

layers. Those layers also contained 53 charcoal pieces and carbonized seeds radiocarbon dated to 13,742 to 11,958 (recently recalibrated with SH20.14c) years before present (cal year B.P.[1]). Optically stimulated luminescence (OSL) and thermoluminescence (TL) analysis confirmed that the last time those soils were exposed to sunlight was between 16,190 ± 930 and 9530 ± 780 years ago. The lumps of pigment were analyzed by scanning electron microscope (SEM_EDS[2]) and found to be chemically consistent with samples from the painted walls above the excavation (Roosevelt 1999: 376–379; Roosevelt et al. 2002: 26, 1996: 190–192). Additionally, some paintings located low on the wall surface today would have been at a more optimal viewing level in relation to the lower position of the cave floor during the Late Pleistocene.

Roosevelt concluded that the earliest inhabitants of the region were Paleoindian hunter-gatherers (broad-spectrum foragers) who nomadically inhabited the hills seasonally for hundreds of years. They created most of the rock art near the end of

[1] B.P. archaeologically refers to "before present" with "present" being set at A.D. 1950.

[2] SEM-EDS refers to scanning electron microscope-energy dispersive spectroscopy to determine weight-percent ratios of iron and titanium in samples versus a control.

the Ice Age before abandoning the site, which was later settled by other cultures more than 3,000 years later. These new groups may have revered the images, as evidenced by handprints overlapping some of the sky-themed paintings. It is also possible that they restored faded paintings, preserved intact ones, and even added a few of their own. Photography of the paintings at *Caverna da Pedra Pintada* and many of the other rock art sites was later conducted by Brazilian archaeologist, Edithe Pereira (2002). From 2007 to 2011, with the assistance of my Brazilian host family and intermittent assistance from some Brazilian archaeology students, and with some funding from the Fulbright foundation, I conducted archaeological research periodically for four years, which included locating, recording, analyzing, and excavating the grounds near red and yellow ochre pictographic rock art to gather evidence and make inferences about the ancient artists who created the art in the hills of the Monte Alegre rainforest preserves (see Fig. 1.2). Analysis of the materials took another 10 years and two brief returns with documentary film crews in 2016 (Glassman 2018) and 2023 (Graham Hancock—Ancient Apocalypse Season 2, ITN productions released Fall 2024 on Netflix).

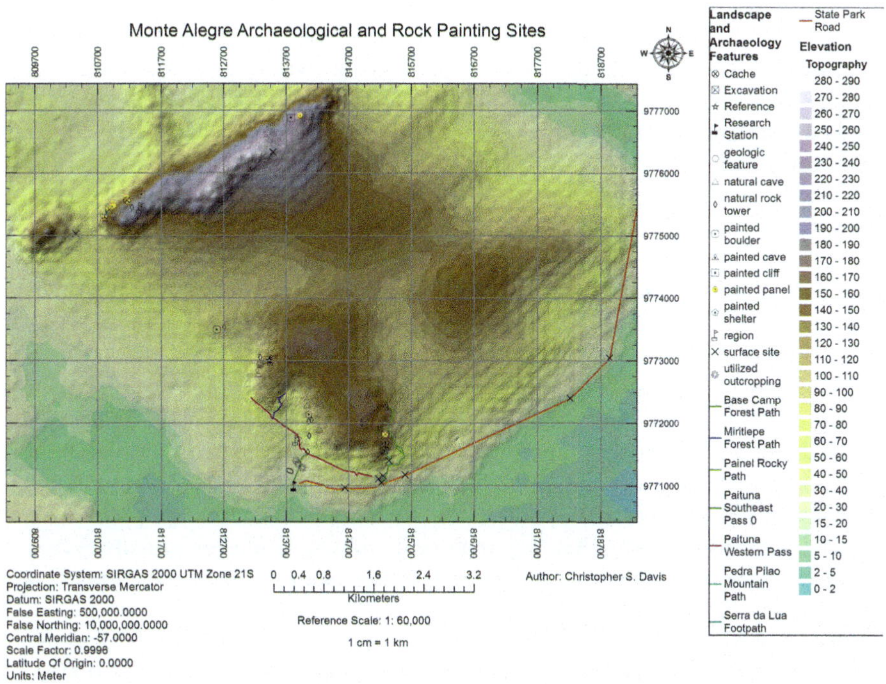

Fig. 1.2 Research Area map showing the three major hills, *Ererê* is the purple diagonal mass at the top of the map, *Maxira* is the brown oval just west of it, *Paituna* is the brownish vertical mass near the lower center of the map. The solid red line curving around from south to east is the road back to the town of Monte Alegre. Archaeological sites, caves, and features are included in the legend. The yellow circles are most of the panels of clustered pictographs

The radiocarbon dates I sent for laboratory analysis from excavation at the *Painel do Pilão* rock shelter (Table 1.1) indicate that the earliest inhabitants came to this region over 13,200 calibrated years before present (B.P., with present professionally defined as A.D. 1950). The *Painel do Pilão* dates are consistent with prior radiocarbon dates taken by Anna Roosevelt's excavations at the nearby *Pedra Pintada* cave just 400 m away. The dates at *Painel* and *Pedra Pintada* similarly cease around 10,800 calibrated years B.P., with no indication of habitation again in the radiocarbon record for over four thousand years.

This research builds upon Roosevelt's and Pereira's works by analyzing and interpreting the paintings for their context and significance. The methods utilized for this task included lithic analysis, iconographic analysis, stratigraphic excavation, robotic photogrammetry, image enhancement and analysis, and retroactive astronomical and sky reconstruction software. I excavated to bedrock in three locations: on top of *Serra do Ererê* hill, in a cave called *15 de Março* on *Serra da Paituna* hill, and at the *Painel do Pilão* shallow rock shelter also on *Paituna* hill. I also recorded and recovered surface collections at eleven different sites, including pottery fragments in *Caverna do Diabo* at the base of *Serra da Maxira* hill and also at some sites in the valley between the hills, turtle bones, and stone tools from *Abrigo dos Ossos 1* on *Serra da Paituna* hill, and red ochre manuports at *Mirante*, another rock art site on *Serra do Ererê*. This book focuses primarily on *Serra da Lua*, *Serra do Sol*, and *Painel do Pilão*.

At *Monte Alegre*, on the *Serra do Ererê* hill, there is a sequence of pictographs on the cliff walls along a relatively narrow footpath, with the cliff mostly facing the western horizon. The westernmost part of the hill, where the sequence begins, is known as *Serra da Lua*. One of the images there depicts a concentric dot that faces the horizon where the winter (southern) solstice sun sets. Similarly, the northwest part of the hill, where the sequence ends, is called *Serra do Sol*, and it features a very large concentric dot that faces the horizon where the summer (northern) solstice sun sets.

Throughout the sequence, there are other concentric circles and dots, including a pictograph that aligns with the setting sun on the equinoxes. Other images interspersed within the sequence depict zoomorphic and anthropomorphic stick figures, as well as disembodied animal and human faces. Several handprints and a few drawn "stick hands" are also found within the sequence. Some handprints overlap drawn images, suggesting that they may be commemorations or acts of reverence. Many other paintings depict a variety of animals, including manatees, turtles, fish, birds, anteaters, monkeys, and humans.

One panel, in particular, may represent an extremely rare and unusual astronomical event: a comet's daytime appearance during a solar eclipse over 13,000 years ago. Because all the paintings face the western horizon and are confined within the boundaries defined by the images aligned with the solstice sunsets, I believe that the other more figurative and abstract paintings within these boundaries function as an ecological "almanac." They likely correlate the time of year with the prominence of certain animals, human activities, and environmental resource cycles, all in sync with the appearance of specific stellar events.

Table 1.1 ^{14}C radiocarbon dates from *painel do pilão* excavation

Sample ID	Exc Level	Depth from surface (cm)	Mass (mg)	14C age BP	d14C age	Ox Cal SH20.14c (INT Cal 20) 95.4% BP			Cal year BP	BC/AD Calibrated Date
20110102-1	2	~18	1091.1	1180	±30	1079–957 (90%)	1178–1165 (3.3%)	1154–1138 (2.2%)	1178–957	A.D. 993–772
20110104-12	4	~60	525.3	3740	±30	4153–3959 (89.5%)	3953–3920 (5.8%)	4216–4214 (0.2%)	4216–3920	2266–1970 B.C.
20110105-20	5	~72	1635.4	5660	±30				6486–6310	4536–4360 B.C.
20110106-17	6	~87	1.60	11,212	±69	13,236–12,967 (92.7%)	12,952–12,927 (2.7%)		13,236–12,927	11,286–10,977 B.C.
20110107-2	7	~90	0.40	10,993	±62	13,020–12,755 (90.5%)	13,069–13,036 (5.0%)		13,069–12,755	11,119–10,805 B.C.
20110107-13	7	~82	1.43	11,088	±64	13,100–12,835 (94.1%)	12,806–12,790 (1.4%)		13,100–12,790	11,150–10,840 B.C.
20110107-14	7	~87	1.97	11,148	±71	13,170–12,895 (93.6%)	12,872–12,850 (1.9%)		13,170–12,850	11,220–10,900 B.C.
20110108-5	8	108	0.07	>30,300						

All carbon excavated by the author, Christopher Davis, in 2011, and sent for accelerated mass spectrometry analysis (the bottom 5 samples) at the NSF-Arizona AMS Laboratory, University of Arizona in 2012 and 2013, and to Beta Analytic in 2019 (top 3 samples). Most recent calibrations were performed with SH20.14c at OxCal_online in 2023

Theory and Significance

Although the constellations perceived by ancient peoples 13,000 years ago are unknowable, the argument made in this book is that there is evidence suggesting attempts to track the movement of the sun throughout the solar year. The division of this span may have been assigned to prominent stellar asterisms, rising cyclically in the cardinal directions throughout the year. In making this argument, I am implying that some ancient cultures created constellations not through pareidolia, but as symbols to represent specific subjects or events tied to particular times of the year—a mnemonic device.

For example, the constellation Leo, resembling a lion, was first recorded around 4000 B.C. in Mesopotamia, coinciding with the summer months, specifically the summer solstice (May 2023; White 2014: 78). By the 3rd millennium B.C., astrology was also being written about, and Leo was assigned to the period from about July 23rd to August 22nd, coinciding with the position of the constellation at that time (Lea 2022). During this period, a royal lion hunt was carved into basalt in Assyria (Brereton 2019). Lion hunts were reserved for chiefs or royalty in Assyria, Egypt, Babylon, Greece, and sub-Saharan Africa thereafter. These hunts took place during the same summer months—approximately between May and October—an interval that remains the legal hunting season in seven African nations today (Loveridge et al. 2009).

Ironically, Leo is no longer visible from August 10th to September 16th in the northern hemisphere night sky, but it was precisely during this time that the royal practice of lion hunting began. The constellation marked the time of year for this human custom. The shape—now recognized as the "sickle" stellar asterism—could have been imagined as any number of objects or animals crouched with a bent neck. Why Leo was chosen for this time of year, Taurus for spring, and Scorpius for autumn all reflect cultural associations practiced in the Levant region during the 3rd and 2nd millennia B.C. Although much of the world later adopted the Levantine zodiac, its original cultural meanings are as lost to us today as the patterns the stars supposedly formed—and even the times of year when they appeared in the night sky.

The ingenuity of associating seasonal activities with star configurations did not begin in Mesopotamia. I believe it was already being practiced by unrelated peoples in the Amazon 10,000 years earlier, based on the evidence at Serra da Lua. This practice was likely common to many unrelated cultures throughout the world, dating back deep into the Ice Age.

My excavations at the open-air sight of *Painel do Pilão* unearthed evidence of wall alteration, buried rocks with painted surfaces, and raw ochre manuports, all of which were radiocarbon dated within a span of time within ~2500 years. The dates end approximately when the Younger Dryas begins (12,850–11,600 calibrated years B.P., which is 10,900–9650 B.C.)—a short climate period defined by sudden and significant global cooling of temperatures to glacial conditions based on buried macrofossil and pollen evidence of the alpine and arctic plant, *Dryas octopetala* throughout the Northern hemisphere. Likewise, some pictographs at *Painel do Pilão* appear to be aligned to solstice and equinox sunrises, and gouges in an altered wall there were possibly made to preserve where sun rays progressed back and forth toward the solstices.

Analysis of the two sites combined has yielded evidence suggesting that Paleoindians at Monte Alegre created one of the earliest solar observatories currently known in the world by painting representative sky phenomena at the sight of observation. The evidence heavily focuses on the sun as the main subject. Whether this was done as a general method for orienting the pioneer inhabitants to a new landscape, or if this was done in recognition of changing environmental conditions remains unsolved. But it attests to the knowledge and ingenuity of humanity during the terminal Pleistocene, before they would have faced the Younger Dryas and the Holocene climate transitions. Did ancient people have, or gain, some inclination of forthcoming climate change, or were they trying to understand change as it began to happen around them?

This book explores a potential shift in epistemological and artistic practices near the end of the Ice Age, a shift that may have granted—or made use of—prior knowledge among Paleoindians in understanding the cyclic movements of the sun, the moon, and the passage of time as they settled into a new ecological region. Research on the ethnoastronomy of recent equatorial South American cultures (Fabian 1992, 2001; Hugh-Jones 1979, 1982; Kelley and Milone 2011; Roe 1996) supports the idea of a shared, deep history of astronomical knowledge, cryptically embedded in some myths and architectural traditions. These cultural repositories combine a wealth of knowledge to preserve information about the annual appearances of constellations and asterisms, and their connections to ecological resources and cultural ceremonies.

More broadly, this research addresses a significant epistemological question: Was there a period in history when art and notation became formal vehicles for learning? Was art initially a medium for emotive and aesthetic expression, with notation added later to convey more practical knowledge? Or were both forms of expression present early in artistic production, with one of their many purposes being the transmission of knowledge to others? The earliest rock art allows time to "pause." By artistically recreating a viewed scene, the moment is frozen, enabling it to be referenced, revisited, and dwelt upon by others later. When did our ancestors first grasp this concept? Such use of imagery laid the foundation for more careful observations and even the consideration of static or dynamic changes over time. Thus, we may find in rock art from around the world a record of sorts that conveys the passage of time. The appearance of painted handprints or hand stencils, for instance, may subtly signify reverence for earlier art.

Pleistocene rock art in Western Europe is some of the oldest rock art in the world, and it has been argued to be representative, symbolic, and aesthetic. It possibly conveys lots of information about the past. But, theories about why it was made revolve around functionalist notions of ritual hunting magic (Breuil 1952)—the concept that drawing a successful hunt will make it really happen for the hunter-gatherers through *sympathetic* (*imitative*) magic.

The remote location of many rock art sites, especially in and near the Pyrenees mountains of Spain and France, precluded simpler theories about the art being made just for representation—or as ad hoc doodles.

Other theories put forth in consideration of a shamanistic perspective (only indisputable for more recent cultures) were that Ice Age art was created to contact and receive benefits from an animistic force (animism/animatism religion), or that it represented landscapes and entities experienced in the spirit world during meditative or hallucinogenic trance (Whitley 2009). Still other theories range from portals or primeval wombs that birthed animals from the underworld (Ross and Davidson 2006), to entoptics, phosphenes, or afterimages formed from optical aberrations (David and McNiven 2019).

The sun imagery at Monte Alegre addresses a different purpose. It marks locations that can be continually consulted to inform not just the initial artist about the sun's position in an annual cycle, but also anyone else educated in its significance. Even if the art was created for personal reasons, its function allows for the education of many. This crossing of art into the threshold of functional education might have occurred sometime near the Terminal Pleistocene about 13,000 years ago. This coincides with other human thresholds too—domestication of plants and animals, portable containers, architectural construction—that might have required continuity of knowledge, and more intensive learning methods for an entire community.The rock art of Monte Alegre suggests that repetitive practices, such as recording sky events by drawing them at a consistent location, may have led to a rudimentary form of observation-based "science." This shift from ritual to process might have sparked the development of an astronomy tradition that linked environmental knowledge to a timeline of cyclical events—a form of pattern recognition. Anthony Aveni (1989: 77) poses the question: "Isn't there a bit of science in a ritual conducted at a time of year signaled by the passage of sunlight down an avenue or through a stone archway? Not a precise science like ours, but nonetheless a predictive empirical science in the sense that one could formally anticipate the time to conduct the ceremony by tracking the position of sunrise."

I agree with Aveni that the foundation of science lies in the detection of patterns (pattern recognition) through repeated observations. By limiting stellar observations to a specific location at a specific time, one can isolate remaining factors and identify correlations. Indeed, ancient artists recording daily observations along the same horizon at sunrise or sunset laid the groundwork for the pattern recognition that we now take for granted as the solar year.

Where cultures associated ecological patterns with celestial events, they could better regulate and even predict the recurrence of important resources—drinkable and navigable waters, natural plant harvests, and animal migrations. A knowledge system of this magnitude could give rise to the Maya calendar, the Egyptian and Mesopotamian cosmologies of the stars, the navigation skills of Polynesians, and the prompt cycles of germination and harvest in agrarian societies. However, Monte Alegre demonstrates that a complex society is not required to develop a sophisticated knowledge of astronomy. Instead, a small or low-density population with the ingenuity and leisure to record observations can create a rich astro-ecological database for future generations to build upon.

Above all, the ability to record knowledge—whether through art or script—allows for the transmission of information across generations. Is this a fundamental aspect of culture distinct to humans? Is this an epistemological revolution that assisted the transition from Upper Paleolithic to Neolithic? From Pleistocene to Holocene? This book posits that it was.

The significance of the rock art at *Monte Alegre* is not in how closely it aligns with the sun and stars, but rather in the fact that it appears to be a record attempting to track their movements. This record is captured in the rock art as notational symbols, not as aesthetic or expressive representations. Perhaps this explains the simplicity of the art. In contrast, the Pleistocene rock art of the Pyrenees region in Europe shows artistic representations that strive for greater realism. Many animals appear lifelike, and Venus figurines or vulva forms are representative efforts. Here, however, humans and animals are depicted as stick figures. While artistic mastery requires practice, if the goal had been realism, the rock artists at *Monte Alegre* had upwards of 2500 years to refine their drawings. If, however, the goal was notational, there may have been no need to improve beyond stick figures.

Some of the earliest writings in Mesopotamia were possibly *bullae*—sealed clay balls containing symbols likely used to indicate items sent for trade (Schmandt-Besserat 2010; Woods 2010). The earliest writings in Egypt were perhaps identification tags for various temple oils and aromatics (Mattessich 2002; Meza 2012). In China, early writing may have taken the form of symbols for divination, written on ox scapulae (Keightley 2006; Mueggler 2022). In each of these cases, writing began as notational symbols, yet they appeared 6000–9000 years after the rock art at *Monte Alegre*. This is not meant to claim that this rock art represents the earliest form of writing, as there is no clear script here. Rather, it is to propose that the goal of the rock art at *Monte Alegre* was to record repeated observations of the sun and some stars, which helped predict future, consistent events—patterns that remain verifiable to this day. This marks an early form of scientific method during a time we often assume was dominated by superstitious animism and magic. This rock art suggests that both may have coexisted.

References

Aveni, A. F. 1989. *Empires of Time: Calendars, Clocks, and Cultures*. New York: Basic Books.

Brereton, G. 2019. Lion Hunting: The Sport of Kings. In *Ancient Mesopotamia*. vol. 2024. London: British Museum.

Breuil, A. H. 1952. *Four Hundred Centuries of Cave Art*. (No Title).

David, B., and I. J. McNiven 2019. *The Oxford Handbook of the Archaeology and Anthropology of Rock Art*, ed. B. David and I. J. McNiven. Oxford University Press.

Fabian, S. M. 1992. *Space-Time of the Bororo of Brazil*. Gainesville: University Presses of Florida.

Fabian, S. M. 2001. *Patterns in the Sky: An Introduction to Ethnoastronomy*. Prospect Heights, Illinois: Waveland Press, Inc.

Glassman, G. 2018. From Caves to Cosmos. In *Native America*, ed. S. Tiffany, pp. 54 minutes. PBS ed. vol. Episode 1, J. Brannum, G. Glassman, J. C. Sousa, S. Tiffany and R. Tinworth, general editor. Providence Pictures.

Hugh-Jones, S. 1979. *The Palm and the Pleiades: Initiation and Cosmology in Northwest Amazonia*. Cambridge, MA: Cambridge University Press.

Hugh-Jones, S. 1982. The Pleiades and Scorpius in Barasana Cosmology. In *Ethnoastronomy and Archaeoastronomy in the American Tropics*, ed. A. F. Aveni and G. Urton, 183–201. New York: Annals of the New York Academy of Sciences New York Academy of Sciences.

Keightley, D. N. 2006. Marks and Labels: Early Writing in Neolithic and Shang China. *Archaeology of Asia* 177–201.

Kelley, D., and E. Milone. 2011. *Exploring Ancient Skies: A Survey of Ancient and Cultural Astronomy*. 2nd ed. New York: Springer.

Lea, R. 2022. *Leo Constellation: Facts, Location, and Stars of the Lion*, ed. Space.com. vol. 2024. New York: Space.com.

Loveridge, A. J., C. Packer, and A. Dutton. 2009. Science and the Recreational Hunting of Lions. *Recreational Hunting, Conservation and Rural Livelihoods: Science and Practice* 108–124.

Mattessich, R. 2002. The Oldest Writings, and Inventory Tags of Egypt. *Accounting Historians Journal* 29 (1): 195–208.

May, N. N. 2023. Towards the History of Representation of the Zodiacal Constellations in Mesopotamia: The Babylonian Man in the Moon and Other Matters. *Archiv für Orientforschung* 42 (43): 208–209.

Meza, A. 2012. *Ancient Egypt before Writing: From Markings to Hieroglyphs*. Xlibris Corporation.

Milhomens, A., D. S. Nascimento, L. Salvo, W. Rodrigues, F. Ferreira, A. Mota, R. Gangana, E. R. de Carvalho, and R. d. C. M. Moreira. 2008. In *Plano De Manejo Do Parque Estadual Monte Alegre (Pema)*, ed. P.-P. d. D. d. E. n. A. Legal, 303. Belem: Ministerio do Meio Ambiente.

Mueggler, E. 2022. Divination and the Scapular Theory of Writing in North Yunnan: Graph, Verse, Page. *Études mongoles et sibériennes, centrasiatiques et tibétaines* (53).

Pereira, E. 2002. *Arte Rupestre Na Amazonia Para*. Belem: Museu Paraense Emilio Goeldi.

Roe, P. G. 1996. Mythic Substitution and the Stars: Aspects of Shipibo and Quechua Ethnoastronomy Compared. *Archaeoastronomy* 12:193–228.

Roosevelt, A. C. 1999. The Development of Prehistoric Complex Societies: Amazonia, a Tropical Forest. *Archeological Papers of the American Anthropological Association* 9 (1): 13–33.

Roosevelt, A. C., J. Douglas, and L. Brown. 2002. The Migrations and Adaptations of the First Americans: Clovis and Preclovis Viewed from South America. In *The First Americans: The Pleistocene Colonization of the New World*, ed. N. G. Jablonski, vol. 27. continuing vols, 159–236. Berkeley: Memoirs of the California Academy of Sciences.

Roosevelt, A. C., M. Lima da Costa, C. L. Machado, M. Michab, N. Mercier, H. Valladas, J. Feathers, W. Barnett, M. Imazio, A. Henderson, J. Silva, B. Chernoff, D. S. Reese, J. A. Holman, N. Toth, and K. Schick. 1996. Paleoindian Cave Dwellers in the Amazon: The Peopling of the Americas. *Science* 272 (5260): 373–384.

Ross, J., and I. Davidson. 2006. Rock Art and Ritual: An Archaeological Analysis of Rock Art in Arid Central Australia. *Journal of Archaeological Method and Theory* 13 (4): 304–340.

Schmandt-Besserat, D. 2010. *How Writing Came About*. University of Texas Press.

White, G. 2014. *Babylonian Star-Lore: An Illustrated Guide to the Stars and Constellations of Ancient Babylonia*. (No Title).

Whitley, D. S. 2009. *Cave Paintings and the Human Spirit: The Origin of Creativity and Belief*. Amherst, New York: Prometheus Books.

Woods, C. 2010. *The Earliest Mesopotamian Writing*. na.

Chapter 2
Geology and Climate Since the Last Ice Age

Geology

About two degrees south of the equator, 680 km upriver from the mouth of the Amazon, the *Gurupatuba* tributary breaks off to the west as the main Amazon River channel bends south. On the northern bank just west of this confluence, the modern-day city of Monte Alegre is perched atop a hill of the same name that forms the southeastern edge of a geomorphological feature called the "Dome of Monte Alegre" (Milhomens et al. 2008: 48). The geomorphology of the dome forms the shape of an oval (15 × 25 km diameters—see Fig. 2.1) with asymmetrical outcrops of hill ridges and columns that are said to resemble "hogbacks." The various summits peak from 50 m to more than 400 m above sea level—all covered with vegetation. "The Dome of Monte Alegre" is considered a structure without equal in the lower Amazon (Milhomens et al. 2008: 48; Vizeu and Almeida 2006).

A tectonic northeast-southwest fault, called the *Ererê Fault*, intrudes the bedrock of the dome, separating the older Paleozoic bedrock into a northern block, which raises *Serra do Itauajarí* to the highest elevation in the region, and a southern block that isolates *Serra da Paituna*. The *Ererê Fault* introduces Tertiary sediments of the *Alter do Chão Formation* (Milhomens et al. 2008: 51; Vizeu and Almeida 2006). This graben-like landmass lies beneath *Serra do Ererê* and the town of *Monte Alegre*. Sulfurous thermal springs also exist in this block, providing evidence that geothermal dynamics remain active in the region.

The *Ererê Fault* passes through part of the Dome of *Monte Alegre* and *Serra do Ererê*, which contains the second tallest peak in the dome. Charles Hartt (1896: 266) was the first scholar to identify a "plexus" of dykes as intrusions of igneous diorite rock. These dykes interrupt the plains, cutting through horizontal layers of schist, clayey schist, and very fine-grained quartzite. Two low valley regions near the *Ererê Fault*—*Desterro* and *Menino Deus*—contain jagged ridges of igneous black rocks. The sulfurous thermal springs, *Serra do Ererê*, and the hot, ammonia-gas-filled

C. S. Davis, *Before Sunset*, Conflict, Environment, and Social Complexity,
https://doi.org/10.1007/978-3-031-93373-8_2

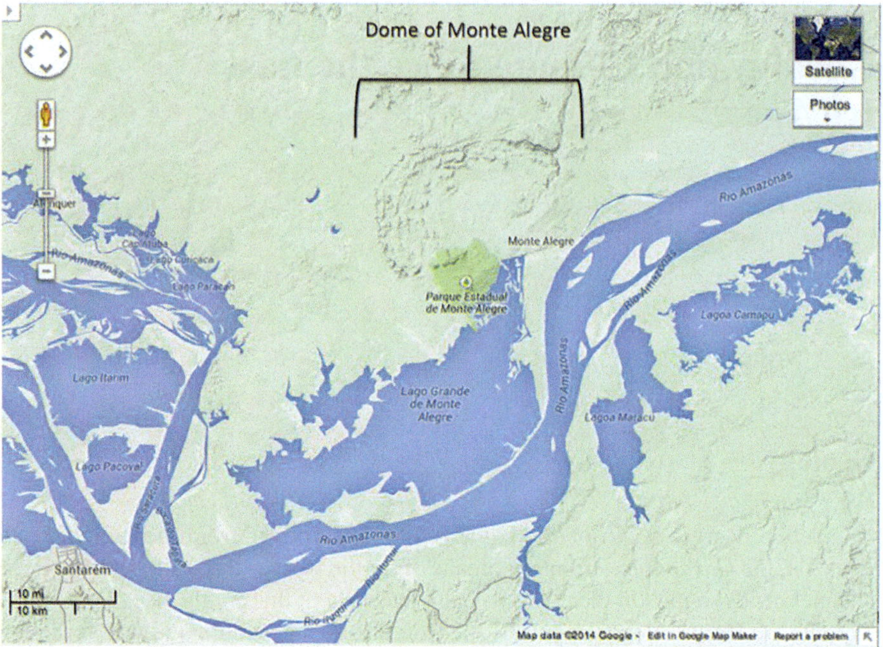

Fig. 2.1 Dome of Monte Alegre relief map, attribution: Map data ©2014 Google, Digital Globe (Google Earth 2014)

Caverna do Diabo cave on *Serra da Maxirá* all lie along the northern contact edge of the *Ererê Fault*.

Evidence of ancient wildfires in sediment layers from the Tertiary and earlier layers were [trans. from Portuguese] “provoked by natural factors, such as: volcanism, electricity discharges, spontaneous combustion, and friction between rocks” (Milhomens et al. 2008: 190). Today, wildfires occur principally in *Serra do Ererê* during the summer period in the *cerrado* biome, where vegetation is most “open” and at its driest. These fires occur [trans. from Portuguese] “when hunting activities intensify by urban groups that camp in the hills, which also coincides with slash-and-burn activities by local residents” (Milhomens et al. 2008: 190).

The rock art sites of *Monte Alegre* are found on both geological strata mentioned previously: the older Paleozoic bedrock (known as the *Ererê Formation*), where *Serra do Ererê* hill rests, and the Tertiary-period *Alter do Chão* sediment, where *Serra da Paituna* hill is located. The *Ererê Formation* was formed during the middle Devonian period and is primarily composed of silt, sandstone, and shale. The lowest levels of this stratum typically consist of gray-colored, silicified, pyritic, mica-rich siltstone, which is sometimes darkened to blue or green, hard, and very friable. Upper stratum levels often contain conglomerates of siltstone with fine layers of well-laminated, dark-gray, hard, mica-rich shale (Milhomens et al. 2008: 114).

Historical reports of fossil discoveries further attest to fossilization conditions within some of the hill microclimates (Milhomens et al. 2008: 51). These findings demonstrate that, despite the rainforest climate—which many scholars assume would only promote decay—hill microclimates can create stable preservation zones for extended periods. While visiting the hills, Charles Hartt (1896: 266) discovered fossils, fragments of fossilized wood, and several impressions of leaves identified by John William Dawson (Milhomens et al. 2008: 51) as *Dicotyledons* from the Tertiary period. According to Dawson, fossils are most likely to be found in fine clay layers atop the spine of *Serra do Ererê* and in the northern portion of *Serra da Paituna*, near the source waters of *Lake Miritiepe*.

The Soils

The valleys between hills in the dome consist of hydromorphic laterites and clays. Rain runoff from the hills creates several temporary streams throughout the valleys during the wet season. Floods from the Amazon River also advance and retreat annually from the south, creating swamplands and marshes from about January through August in the Lages community, east of *Serra da Paituna*. These floods also saturate the soil around *Lake Miritiepe*, located slightly northwest of Lages (Milhomens et al. 2008: 48; Oliveira Junior et al. 1999). The oval valley at the center of the dome contains the most sediment from erosion. This central valley, located north of the community of *Ererê*, has soil profiles ranging from 10 to 50 m in depth. Some of these soils form large grass-covered mounds of latosols that vary in color from yellow (under dystrophic conditions) to red (Milhomens et al. 2008: 114). Today, many of these mounds are utilized for agriculture.

Soil profiles at many rock art sites primarily consist of dystrophic reddish-yellow podzol rocky soils with medium to coarse-grained clay. These are often mixed with desiccated cambisols, influenced by their proximity to some of the outcrops in these hill ranges. This mixed soil type is most common on slopes with a declination of between eight and forty-five percent. These soils are classified as sandy and hydromorphic, providing good, though not rapid, drainage. In the low-lying floodplains, soils are predominantly hydromorphic laterites and gley. Conversely, alluvium in the riverbeds and riverbanks forms poorly developed soils but is always rich in organic material (Milhomens et al. 2008: 115).

The Hill Ridges

The southernmost hill ridge, *Serra da Paituna*, lies closest to the Amazon River. It is oriented north-south, with two round peaks and abrupt scarps on its western flanks (Milhomens et al. 2008: 48). This hill ridge spans about 3 km in length and reaches approximately 200 m in altitude. The most notable feature in this hill range is *Pedra do Pilão*[1]—a pillar of friable sandstone jutting 20 m high on the southern flank, and capped with a pseudo-karstic[2] tabular block (Milhomens et al. 2008: 114–115) almost resembling a flat arrow, or at least a "T" shape (Fig. 2.2). The tip of the "arrow" faces ~18° North, and the two back corners face ~216° and ~237°.

On the southern flank of *Serra da Paituna*, several other rock art sites are positioned inside several caves, grottoes, and rock shelters. Only one other rock art site on *Paituna* besides *Pedra do Pilão* can be described as "open-air" or "outdoors"—*Painel do Pilão*.

There, its many surfaces at two different levels span cardinal directions from roughly southwest to northwest (the entire eastern and northern sky, which includes all sunrises throughout the year), and its pointed-rim overhang seems to have protected its painted surface from strong erosive forces for a very long time despite weather flowing from east to west. Rain seems to rarely, if ever, touch its painted surface.

About 2 km north, and oriented nearly perpendicular (east-west) to the *Paituna* hill ridge is *Serra do Ererê* (Fig. 2.3). This hill ridge is longer and taller, spanning about 4 km in length by one-and-a-half to two kilometers in width, and reaching 255 m high. It has steep scarps on the north and south flanks, and abrupt slopes or precipices, sometimes more than 100 m high. Some sheer walls there have fractures that form crevices along their entire length, while a few slopes terminate in numerous disparate boulders and sandy soil (Milhomens et al. 2008: 114).

Most pictographs on Ererê are found on its western flank, although there is one known rock art site on its eastern flank, called Mirante, and a cave containing a few paintings on its southern flank. The pictographs on the western flank have traditionally been divided between two distinct sites: Serra da Lua on the westernmost edge of Ererê and Serra do Sol on its northwest flank. This division is due to forest regrowth and steeply sloping terrain that now separate the two sites, disrupting any footpaths that may have connected them in the past.

Fig. 2.2 Photo of *Serra da Paituna* (right) as seen from a boat in the Amazon River during the flood season. The small "T"-shape at the top of the hill is *Pedra do Pilão* (left)

[1]According to Katzer (1933), *Pedra do Pilão* also went by the names of "*induá-mena*" and "*mão de pilão*" (Milhomens et al. 2008: 51, 114). [*induá-mena* is Tupi for *pillar-male animal*; *mão de pilão* is Portuguese for *hand of the pillar*]

[2]Possibly a quartzite "tor".

Fig. 2.3 *Serra do Ererê* as seen from the Amazon River during the flood season. The small "knob" at the top of the hill on the right side of the photo is *Mirante*

Mirante, located on the eastern flank of Ererê, is a massive stand-alone outcrop at the hill's eastern peak. However, all the pictographs at the site are found on the western side of this outcrop, painted on its sheer flat wall. Like the pictographs on the western flank of Ererê, the eastern rock art also faces west. On the southwest edge of the Mirante outcrop, there is a path that can be scaled to reach the top, offering a 360° view of the region. This site is called Mirante, meaning "lookout" or "view."

The cave on the southern flank of Ererê is called Itatupaoca, with its wide, 9.5-m opening divided by a single horizontal bar about 1.6 m from the ground and a zone of friable horizontal layers approximately 4.5 m above ground (Milhomens et al. 2008: 117–118). Its name comes from the Tupi language family and translates literally to "stone, father, house" (Katzer 1933). The meaning could refer to a "grand home," "cave house," or possibly "dwelling of the spiritual father," "stone house of God," or "stone church." The only visible art associated with the cave is on the ceiling outside its entrance. Despite the presence of the cave and historical references to it containing art in the past, all recorded rock art on Ererê is open-air, nearly all of which face west or northwest.

Just west of *Ererê* is *Serra da Maxirá* (Fig. 2.4). With its round base and an altitude of only about 150 m (Milhomens et al. 2008: 114), it more appropriately fits the description of a hill compared to the other two ridges. There is one known cave art site in Maxirá, located in an ammonia-gas-filled cave called *Caverna do Diabo*. The cave entrance is situated at the base of the hill, near the center of its southern flank. On the hillside above the cave, there are several large white boulders, each with at least one blackened side. This entire area feels hotter than anywhere else on

Fig. 2.4 *Serra da Maxira*. The third hill ridge as seen from a road on its west flank (with Ererê in the distance to the right). *Maxira* is located west of *Ererê* and northwest of *Paituna*. The local population names the hill for the community of *Maxira* nearby, and signs have been placed along the dirt roads near this hill bearing the name. *Maxira* is labeled in some books as *Serra do Bode*, which means hill of the butterfly in Portuguese

the hill due to hot gases that penetrate the cave and its roof (Derby 1898; Milhomens et al. 2008: 117). The ground also dips lower than its immediate surroundings, which is likely why this landscape feature has been called *"buraco quente"*—meaning "hot hole," or more colloquially, "hell hole."

The hill of Maxirá is also significant to the rock art on *Ererê* because it stands directly in front of the western horizon of some paintings at *Serra da Lua*. Ethnographic studies of more recent circum-Caribbean cultures indicate that some groups refer to the land in the west, or particularly behind mountains, as the "land of the dead" (Guchte and Brewer 1992: 20; Léry 1992: 91, 136; Southey 1810: 247–248). Based on these ethnographic analogies, the paintings in *Caverna do Diabo* might have death-themed or ancestral-spirit significance.

Presently, the area is not known to have major earthquakes, volcanic eruptions, damaging winds, or tempestuous storms. There is seasonal flooding, but no evidence suggests that any catastrophic floods washed away large tracts of land or gouged canyons out of the terrain. In fact, the current environment suggests that Monte Alegre has remained relatively stable for a very long time. Sun, precipitation, and humidity can be intense at times, and surely, the natural landscape has evolved in response to them. However, the very fact that Amazon-specific lichen grows in the hills here[3] is proof that the climate has been consistent for quite some time, as similar hill lichens are indicators of "pristine" environments (Jovan 2008).

Current Climate

Today, *Monte Alegre* receives an average of 1780 mm of annual rainfall, balancing between a rainy and a dry season. Average annual temperatures remain above 22.5 °C, never dropping below 18 °C. This designates the climate as *Am* (Tropical Monsoon) according to the updated Köppen climate classification (Getis et al. 2010). The rainy season begins in December and lasts until June, with rainfall gradually increasing until March. Monthly precipitation can exceed 200 mm, with the wettest months occurring from March to May. April typically sees the highest rainfall amounts, averaging 317.2 mm. The "dry" season extends from July to November, with the driest months occurring between September and November, when rainfall amounts can drop to as low as 60 mm in a given month (Milhomens et al. 2008: 48, 103; Pastana 1999; Vizeu and Almeida 2006).

The dry months are also the windiest, with wind speeds reaching up to 3.9 m/s, while they dip to around 2.0 m/s during February and March. Winds blow predominantly from east to west throughout the year, with local variations influenced by elevated relief. Air pressure peaks in July at approximately 997 mm Hg and reaches its minimum in mid-December, averaging around 992.5 mm Hg. Relative air humidity closely follows the annual precipitation cycle, fluctuating from 62.8% in October to 79.9% in March and April, with an overall annual average of 72.8% (Milhomens et al. 2008: 103; Oliveira Junior et al. 1999).

Pleistocene Ice Ages

To many, the phrase *ice age* fills the imagination with icebergs, igloos, and polar bears.

However, during an *ice age*—like now—some places on Earth remained warm, others remained cold, and most places fluctuated seasonally between warm and cold. More accurately, the Pleistocene—which is the geological epoch that spanned from approximately 2.6 million to 11,700 years ago—contained several "ice ages."

[3] See results section for *Laurera purpurina* and *Chrysothrix xanthina*; also green lichen is *Cladonia coccifera*—its red "fruit" makes dye (see Zahlbruckner 1909: 261). Jovan (2008: 34–38) explains that while precipitation and temperature are common factors affecting lichen growth, nitrogen exposure can affect lichen growth through two pathways: direct air respiration of the lichen, and through alteration of bark pH where the lichen grows on trees. Increased nitrogen can promote the growth of nitrophytic lichen, but it can also weaken neutrophytic lichen and destroy cyanolichen and acidophytic lichen.

Continental glaciers (ice sheets) spread to latitudes lower than 65°, and alpine snowlines descended below their average elevations today of 4900 m (16,000 feet) above sea level (asl) (Zimmermann 2017).

Weather, in general, is Earth's movement of heat from greater to lesser concentrated areas. As the sun shines on this rotating planet, some areas warm greater than others, causing high and low air pressure zones, winds and jet streams, precipitation, and fluctuating temperatures. Climate refers to longer-term weather trends in specific regions on earth, and the Pleistocene was a period when vast regions of Earth's higher latitudes were more cool than hot, allowing for accumulation of snow and ice year after year, for long periods of time. However, the Pleistocene also contained several warmer periods, called interstadials; its final one is called the Bølling-Allerød warm period.

The Bølling-Allerød Warming Period

The *Bølling-Allerød* period refers to ~14,690 to ~12,890 years before present (BP), equivalent to the calendar years 12,740 to 10,940 B.C. However, closer scrutiny reveals that this period is further divided into two distinct warming phases: the *Bølling* (14,650–14,000 BP) and the *Allerød* (13,600–13,300 BP). These phases were identified by the sudden absence of *Dryas* pollen and the emergence of *Betula* (birch) pollen in a peat layer dredged from *Bølling Lake* in Denmark (Bennike and Mortensen 2018). Extensive evidence supporting these date ranges has since been obtained from other sediment cores, oxygen isotope data from Greenland ice cores, and the spread of northern-latitude arboreal forests (e.g., oak, birch, and pine). This warming also coincided with a ~35-m sea level rise caused by glacial melting (Krüger and Damrath 2020; Mangerud 2021; Nogarotto et al. 2023).

Average global temperature estimates for this warm period are difficult to determine, but temperature changes in Europe have been well simulated using geological, lake level, paleoecological, and atmospheric circulation models. These simulations indicate seasonally differential warming in northwest Europe at the onset of the *Bølling* period. Winter temperatures rose dramatically, from −25 °C during the Oldest *Dryas* to −5 °C during the *Bølling* oscillation (a +20 °C increase), while summer temperatures rose more modestly, from ~10 °C during the Oldest *Dryas* to ~13 °C during the *Bølling* warm phase (Renssen and Isarin 2001). This seasonal differential warming was theorized to result from melting sea ice at the margins of land ice, allowing winter ocean upwellings to warm northern Europe more significantly in winter than in summer.

Outside of Europe, the *Bølling* warming also triggered an abrupt increase in precipitation from summer monsoons in North Africa (Clark et al. 2012). After the ~500-year *Bølling* warm spell, evidence from parts of the world suggests global temperatures returned to cold glacial conditions again, based on the return of the tundra flower, *Dryas octapetala*, for which the period is named the "Older Dryas." This was brief, though, until another warm disruption—the *Allerød* interstadial, dated from ~13,600 to 12,900 ± ~300 years BP (11,650–10,950 B.C.)—was identified within ancient clay deposits near the town of Allerød. Then, *Dryas octopetala* flowers returned again, as before, defining the period called the Younger Dryas—which is dated by multiple lines of evidence to span from 12,807 ± 12 to ~11,700 ± 40 calibrated years BP (10,900 – 9700 B.C.) (Cheng et al. 2020; Rasmussen et al. 2006; Reinig et al. 2021).

The Holocene epoch began next, which is our current period of warmer global temperatures that have bathed the rise of human civilizations and scientific technology. The total time span, from the *Bølling* warming to the beginning of the Holocene, is five temperature reversals from cold (oldest Dryas) to warm (*Bølling*) to cold (Older Dryas) to warm (*Allerød*) to cold (Younger Dryas) to warm (Holocene)—a span of ~3,000 years (14,690–11,700 cal BP)!

The source of Pleistocene climate change has long been debated. One longstanding theory centers on solar insolation, which refers to the intensity of the sun's warmth as it varies based on three celestial cycles known as the *Milankovitch cycles* (Levy 2020; Szarka et al. 2021). Serbian climatologist Milutin Milankovitch proposed this theory in the 1920s, and it has been widely accepted as a major driver of long-term climate fluctuations. However, many critics argue that shorter-term climate fluctuations are not adequately explained by these generally long-term insolation cycles.

Other theories and variables—such as ice albedo, ocean salinity, deep-water ocean currents and upwellings, geomagnetic deflections, pole reversals, sunspot activity, cosmic impacts, and human-induced effects—have been proposed as alternative or complementary explanations for climate oscillations, particularly during the end of the Pleistocene (Bond et al. 1992). Some, like cosmic impacts, have gained mainstream popularity, while others, such as ocean salinity and circulation, have been favored by specific academic disciplines like geology and archaeology. Most detractors of the Milankovitch theory point to one climatic period as evidence supporting their preferred explanations: the *Younger Dryas*.

The purpose of this chapter is neither to defend nor to reject any specific theory regarding the causes of these climate transitions. Instead, it seeks to comprehend the scope and scale of the changes that occurred around the time ancient people began settling along the Amazon River. To be clear, I am not arguing that changing climate determined cultural responses. Rather, I aim to explore what role, if any, these changes may have played. Was the climate change drastic and rapid enough to be noticed by Paleoindians, thereby spurring a novel response such as tracking the sun through rock art? Or were the changes too gradual or subtle to have been perceived, making them unlikely to have triggered innovative cultural behaviors?

The Younger Dryas

The Younger Dryas was brief in geological history—only lasting around 1100 years, from ~12,800 to 11,700 calibrated years BP (10,900–9700 B.C.), and both its onset and termination were abrupt—with temperature swings occurring in as little as 50, or even 10 years (Cheng et al. 2020; Rasmussen et al. 2006; Reinig et al. 2021). During the Younger Dryas, average global air temperatures are estimated to have dropped by about 7 °C (~12 °F) (Boulton et al. 2004), but polar regions like Greenland have ice-core and oxygen isotope evidence indicating temperatures in those regions dropped by as much as 10 °C (18 °F), while sediment cores from Cariaço Basin off the coast of Venezuela indicate a tropical temperature decline of ~3 °C (~5 °F). This translates to an average global surface temperature of ~12–12.5 °C (53.6–54.5 °F).

For comparison, average global air temperatures for the last glacial maximum were about 11 °C (51.8 °F) and for the twentieth century A.D. they are ~14 °C (57 °F) (Lindsey and Dahlman 2021). The twentieth century average is based on average yearly surface temperatures (of land and ocean) from 1951 to 1980, and climate change from A.D. 1977–2021 has increased above this average by ~1.01 °C (~2 °F)—a span of 45 years (Shaftel 2022). Indeed, if in 10, 50, or even 100 years the Younger Dryas underwent a temperature drop of 3 °C (equatorial regions) to 10 °C (polar regions)—7 °C average—then it should have been very noticeable to ancient people living through it anywhere on Earth.

Paleoclimate of the Monte Alegre Rock Art Sites

Because the results of this research demonstrate human activity from 13,286 to 11,650 calibrated years B.P., it is important to directly address the climate during this time frame. Both polar ice-core and equatorial sea coral records suggest that this was a critical period when our human ancestors were faced with rapid changes to the global climate.

Landmasses near the equator experienced a total rise in sea level of over 100 m since the Last Glacial Maximum 17,000 years ago to about 7000 years ago (Fairbanks 1989). The lower sea level may have caused an increased drainage gradient, causing the Amazon River and its tributaries to cut tens of meters below their current floodplain levels (Betancourt 2000: 2275).

Since then, the greatest single increase in sea level has been identified as the *Meltwater Pulse 1A* event, occurring around 14,600 to 13,800 years BP. Many scholars attribute this rise to glacial melting during the *Bølling-Allerød* warm interval (1996; Bard et al. 1990; Deschamps et al. 2012; Weaver et al. 2003). Notably, *Monte Alegre* was first inhabited about 500 years after this pulse but about 500 years before the *Younger Dryas*. Were Paleoindian pioneers uprooted from their original homeland in part due to warmer, wetter conditions? Or did these warmer conditions enhance their intrepid sense of exploration, making a deeper and more easily navigable Amazon River more inviting? Perhaps neither, and there is no direct correlation to climate. However, it is noteworthy that the first period of human activity at *Painel do Pilão* and Roosevelt's (1996) identified Paleoindian occupation at *Caverna da Pedra Pintada* both end at about the same time that the *Younger Dryas* period begins.

Paleoclimate data from dissolved noble gases in groundwater, pollen records, snowline reconstructions, strontium/calcium ratios, and $\delta^{18}O$ coral records indicate that during the *Last Glacial Maximum* (~17 kya), Northern Brazil was 5.4° ± 0.6 °C cooler than it is today (Stute et al. 1995). Some scholars interpret this as cooler, drier conditions akin to a savannah-like climate. However, abrupt isotopic changes, coupled with pollen records from sediment cores taken from *Lagoa de Caço* in Maranhão, Northern Brazil, suggest a shift to humid rainforest conditions beginning between 17.3 and 16.8 thousand years ago. These conditions have persisted to the present day, except for a brief period of increased aridity during the *Younger Dryas* event (Jacob et al. 2007).

Similar evidence of wetter conditions is found in sediment cores from *Lagoa de Pata* in the forested Rio Negro Basin of Amazonas, Northern Brazil, which show increased total organic carbon and chlorophyll levels starting around 15.3 thousand calibrated years BP (Cordeiro et al. 2011). However, research by Maslin and Burns (2000) provides a more detailed record of Amazon River discharge, revealing a ~60% reduction from present outflow between 13,000 and 11,800 years BP—corresponding to the *Younger Dryas*. This was followed by a sudden ~40% increase in outflow at the end of the event, attributed to increased lowland rainfall rather than Andean meltwater (Betancourt 2000: 2275; Maslin and Burns 2000: 2286–2287).

Although less specific, similar conclusions are supported by more recent research indicating that a collapsed Atlantic Meridional Overturning Circulation (AMOC), which is believed to have occurred during the period after Meltwater Pulse 1a, causes the tropical rain belt to shift further south, resulting in what should have been a dryer northern Amazon (Barker and Lear 2023: 15). Specific location differences are still uncertain, though, and the northeastern Amazon would probably have seen increased rainfall. Monte Alegre is in the Northern Amazon, but also close to the northeast.

Together, the paleoclimate record suggests that much of the Amazon was cooler and dryer during the Last Glacial maximum >17,000 years ago, then there was a shift to more humid and tropical conditions ~16,000 years ago, an increase of water levels spiked around 15,000 to 14,000 years ago, then a significant drying effect and lowering of Amazon River drainage took place during the Younger Dryas ~12,800–11,700 calibrated years BP (Cheng et al. 2020; Rasmussen et al. 2006; Reinig et al. 2021). This was followed by the continuous Holocene warm, tropical conditions that persist today. Since these concordant paleoclimate records geographically encompass our research area, we can assume that Monte Alegre experienced this trend of at least slightly dryer conditions during much of the first habitation period.

Additional climate details include evidence from n-alkane $\delta^{13}C$ taraxerol and levoglucosan concentrations taken from ocean cores at the Amazon River fan (Maslin et al. 2012), which indicate that vegetation in the Amazon basin during the Younger Dryas remained forested, showing no increase of grassy savannah refuges, despite the increased arid conditions. Further evidence (Cordeiro et al. 2008) indicating the persistence of this biome since the Younger Dryas was determined from paleofire data taken from a lake in Carajás, East Amazonia. Although lake sediment records begin here about 11,800 years ago, it did not contain any significant charcoal sediment from wildfires or biomass burning until the mid-Holocene, around

7450 to 4750 calibrated years before present, and then again from 1300 to 70 calibrated years before present (Cordeiro et al. 2008: 60). These results support the notion that the rainforest biome did not dry to the point where wildfires and savannahs were a common occurrence.

The timeframe of this research reflects largely on the onset of the Younger Dryas event, but the paleoclimate data suggest that conditions were not so drastic that it changed the type of biome from what it is today. Therefore, we can conclude that while some changes to the frequency of certain animal species and the prevalence of particular vegetation might have occurred, the ecosystem has remained relatively stable since the *Bølling-Allerød* warm interval.

The extent to which Paleoindians needed to adapt to the cooler and drier conditions during the *Younger Dryas* remains unresolved, but a review of the climate data suggests they would have recognized some changes. The traditional view of human populations during the Pleistocene period is that they were hunter-gatherers with largely animistic beliefs and keen practices (Bird-David 1990, 1992, 1999; Conkey et al. 1997; Deacon 1989; Helvenston and Hodgson 2010; Mazza et al. 2006). Whether or not they acknowledged and attributed these climate changes to supernatural forces or spirits, research indicates that Paleoindians at *Monte Alegre* were broad-spectrum foragers (Oliver 2008: 185–212; Roosevelt et al. 1996, 2002: 202). They appear to have developed a system of recording careful and systematic observations in paintings, which likely helped them predict cyclic patterns of resource availability in the local environment. Current evidence appears to favor the notion that they developed (or already knew) astronomy-related skills that made their foraging efforts more efficient. The knowledge of orienting oneself to seasonal changes by utilizing astronomy observations might have been crucial for broad-spectrum foragers at the end of the Ice Age to survive this period of climatic transition.

References

Bard, E., B. Hamelin, M. Arnold, L. Montaggioni, G. Cabioch, G. Faure, and F. Rougerie. 1996. Deglacial Sea-Level Record from Tahiti Corals and the Timing of Global Meltwater Discharge. *Nature* 382 (6588): 241–244.

Bard, E., B. Hamelin, and R. G. Fairbanks. 1990. U-Th Ages Obtained by Mass Spectrometry in Corals from Barbados: Sea Level During the Past 130, 000 Years. *Nature* 346 (6283): 456–458.

Barker, S., and C. H. Lear 2023. Climate Tipping Point Interactions and Cascades: A Review. *Earth System Dynamics (Pembroke, Ont.)*

Bennike, O., and M. F. Mortensen. 2018. A Multi-Disciplinary Macrofossil Study of Late Glacial to Early Holocene Sediments from Søndre Kobberdam, Hareskovene, Denmark. *Bulletin of the Geological Society of Denmark* 66:113–122.

Betancourt, J. L. 2000. The Amazon Reveals Its Secrets--Partly. *Science* 290 (5500): 2274–2275.

Bird-David, N. 1990. The Giving Environment: Another Perspective on the Economic System of Gatherer-Hunters. *Current Anthropology* 189–196.

Bird-David, N. 1992. Beyond 'the Hunting and Gathering Mode of Subsistence': Culture-Sensitive Observations on the Nayaka and Other Modern Hunter-Gatherers. *Man* 19–44.

Bird-David, N. 1999. "Animism" Revisited: Personhood, Environment, and Relational Epistemology 1. *Current Anthropology* 40 (S1): S67–S91.

Bond, G., H. Heinrich, W. Broecker, L. Labeyrie, J. McManus, J. Andrews, S. Huon, R. Jantschik, S. Clasen, C. Simet, K. Tedesco, M. Klas, G. Bonani, and S. Ivy. 1992. Evidence for Massive Discharges of Icebergs into the North Atlantic Ocean During the Last Glacial Period. *Nature* 360 (6401): 245–249.

Boulton, G. S., P. Dongelmans, M. Punkari, and M. Broadgate. 2004. Evidence of European Ice Sheet Fluctuation During the Last Glacial Cycle. In *Developments in Quaternary Sciences*, ed. J. Ehlers and P. L. Gibbard, vol. 2, 441–460. Elsevier.

Cheng, H., H. Zhang, C. Spötl, J. Baker, A. Sinha, H. Li, M. Bartolomé, A. Moreno, G. Kathayat, and J. Zhao. 2020. Timing and Structure of the Younger Dryas Event and Its Underlying Climate Dynamics. *Proceedings of the National Academy of Sciences* 117 (38): 23408–23417.

Clark, P. U., J. D. Shakun, P. A. Baker, P. J. Bartlein, S. Brewer, E. Brook, A. E. Carlson, H. Cheng, D. S. Kaufman, Z. Liu, T. M. Marchitto, A. C. Mix, C. Morrill, B. L. Otto-Bliesner, K. Pahnke, J. M. Russell, C. Whitlock, J. F. Adkins, J. L. Blois, J. Clark, S. M. Colman, W. B. Curry, B. P. Flower, F. He, T. C. Johnson, J. Lynch-Stieglitz, V. Markgraf, J. McManus, J. X. Mitrovica, P. I. Moreno, and J. W. Williams. 2012. Global Climate Evolution During the Last Deglaciation. *Proceedings of the National Academy of Sciences* 109 (19): E1134–E1142.

Conkey, M. W., O. Soffer, D. Stratman, and N. G. Jablonski, eds. 1997. *Beyond Art; Pleistocene Image and Symbol*. San Fransico: California Academy of Sciences.

Cordeiro, R. C., B. Turcq, A. Sifeddine, L. D. Lacerda, E. V. S. Filho, B. Gueiros, Y. P. Potty, R. E. Santelli, E. O. Pádua, and S. R. Patchinelam. 2011. Biogeochemical Indicators of Environmental Changes from 50 Ka to 10 Ka in a Humidregion of the Brazilian Amazon. *Palaeogeography, Palaeoclimatology, Palaeoecology* 299:426–436.

Cordeiro, R. C., B. Turcq, K. Suguio, A. O. d. Silva, A. Sifeddine, and C. Volkmer-Ribeiro. 2008. Holocene Fires in East Amazonia (Carajás), New Evidences, Chronology and Relation with Paleoclimate. *Global and Planetary Change* 61:49–62.

Deacon, H. J. 1989. Late Pleistocene Palaeoecology and Archaeology in the Southern Cape, South Africa. In *The Human Revolution: Behavioural and Biological Perspectives of the Origins of Modern Humans*, ed. P. Mellars and C. B. Stringer, 547–564. Scotland: Edinburgh University Press.

Derby, O. A. 1898. A Serra Do Mixixá Em Trabalhos Restantes Inéditos Da Comissão Geológica Do Brasil (1875–1878). *Boletim do Museu Paraense de História Natural e Etnographia* 2 (1–4): 340–343.

Deschamps, P., N. Durand, E. Bard, B. Hamelin, G. Camoin, A. L. Thomas, G. M. Henderson, J. I. Okuno, and Y. Yokoyama. 2012. Ice-Sheet Collapse and Sea-Level Rise at the Bølling Warming 14,600 Years Ago. *Nature* 483:559–564.

Fairbanks, R. G. 1989. A 17,000-Year Glacio-Eustatic Sea Level Record; Influence of Glacial Melting Rates on the Younger Dryas Event and Deep-Ocean Circulation. *Nature* 342 (637–642):

Getis, A., J. Getis, M. Bjelland, and J. Fellmann. 2010. *Introduction to geography*. 13th ed. New York: McGraw-Hill.

Google Earth. 2014. Parque Estadual De Monte Alegre, Lat -2.044724, Lon 54.180469, Elev 132 M. Terrain Map Data Layer. Google, Digital Globe, Google Earth.

Guchte, M. v. d., and D. Brewer (editors). 1992. *Masquerades and Demons: Tukuna Bark-Cloth Printing*. Urbana: Krannert Art Museum—Board of Trustees of the University of Illinois.

Hartt, C. F. 1896. A Geologia Do Para. *Boletim do Museu Paraense Emilio Goeldi de Historia Natural e Etnografia* Tomo 1 (Fasciculos 1-4): 257–273.

Helvenston, P. A., and D. Hodgson. 2010. The Neuropsychology of 'Animism': Implications for Understanding Rock Art. *Rock Art Research* 27 (1): 61–94.

Jacob, J., Y. Huang, J.-R. Disnar, A. Sifeddine, M. Boussafir, A. L. S. Albuquerque, and B. Turcq. 2007. Paleohydrological Changes During the Last Deglaciation in Northern Brazil. *Quaternary Science Reviews* 26:1004–1015.

Jovan, S. 2008. *Lichen Bioindication of Biodiversity, Air Quality, and Climate: Baseline Results from Monitoring in Washington, Oregon, and California*, ed. U. S. D. o. Agriculture. (USDA), Pacific Northwest Research Station.

Katzer, F. 1933. Geologia Do Estado Do Pará. In *Boletim do Museu Paraense Emílio Goeldi de História Natural e Etnografia*, ed. M. P. E. Goeldi. Belém, Brazil.

Krüger, S., and M. Damrath. 2020. In Search of the Bølling-Oscillation: A New High Resolution Pollen Record from the Locus Classicus Lake Bølling, Denmark. *Vegetation History and Archaeobotany* 29:189–211.

Léry, J. d. 1992. *History of a Voyage to the Land of Brazil, Otherwise Called America*. Berkeley, CA: University of California Press.

Levy, R. 2020. Milutin Milankovitch (1879–1959): Orbital Variations. In *NASA Earth Observatory*. NASA Earth Observatory.

Lindsey, R., and L. Dahlman. 2021. Climate Change: Global Temperature. In *NOAA National Centers for Environmental Information, State of the Climate: Global Climate Report for Annual 2021*, ed. J. Blunden, vol. 2022. NOAA Climate.gov.

Mangerud, J. 2021. The Discovery of the Younger Dryas, and Comments on the Current Meaning and Usage of the Term. *Boreas* 50 (1): 1–5.

Maslin, M. A., and S. J. Burns. 2000. Reconstruction of the Amazon Basin Effective Moisture Availability over the Past 14, 000 Years. *Science* 290:2285–2287.

Maslin, M. A., V. Ettwein, C. S. Boot, J. Bendle, and R. D. Pancost. 2012. Amazon Fan Biomarker Evidence Against the Pleistocene Rainforest Refuge Hypothesis? *Journal of Quaternary Science* 27 (5): 451–460.

Mazza, P. P. A., F. Martini, B. Sala, M. Magi, M. P. Colombini, G. Giachi, F. Landucci, C. Lemorini, F. Modugno, and E. Ribechini. 2006. A New Palaeolithic Discovery: Tar-Hafted Stone Tools in a European Mid- Pleistocene Bone-Bearing Bed. *Journal of Archaeological Science* 33:1310–1318.

Milhomens, A., D. S. Nascimento, L. Salvo, W. Rodrigues, F. Ferreira, A. Mota, R. Gangana, E. R. de Carvalho, and R. d. C. M. Moreira 2008. *Plano De Manejo Do Parque Estadual Monte Alegre (Pema)*, ed. P.-P. d. D. d. E. n. A. Legal, 303. Belem: Ministerio do Meio Ambiente.

Nogarotto, A., R. Noormets, T. Chauhan, G. Mollenhauer, J. Hefter, H. Grotheer, S. T. Belt, F. Colleoni, F. Muschitiello, L. Capotondi, C. Pellegrini, and T. Tesi. 2023. Coastal Permafrost Was Massively Eroded During the Bølling-Allerød Warm Period. *Communications Earth & Environment* 4 (1): 350.

Oliveira Junior, R. C. d., T. E. Rodrigues, P. L. Dos Santos, and M. A. Valente 1999. *Zoneamento Agroecológico Do Município De Monte Alegre, Estado Do Pará*, ed. CPATU, 87. Documentos 9 ed. Série Documentos (CPATU). vol. INFOTECA. Belém: EMBRAPA Amazônia Oriental.

Oliver, J. R. 2008. The Archaeology of Agriculture in Ancient Amazonia. In *Handbook of South American Archaeology*, ed. H. Silverman and W. H. Isbell, 185–216. New York: Springer.

Pastana, J. M. d. N. 1999. Diagnóstico Do Potencial Ecoturístico Município De Monte Alegre. In *Programa de Integração Mineral em Municípios da Amazonia (PRIMAZ)*. Belém: Ministério de Minas e Energia.

Rasmussen, S. O., K. K. Andersen, A. M. Svensson, J. P. Steffensen, B. M. Vinther, H. B. Clausen, M. L. Siggaard-Andersen, S. J. Johnsen, L. B. Larsen, D. Dahl-Jensen, M. Bigler, R. Roëthlisberger, H. Fischer, K. Goto-Azuma, M. E. Hansson, and U. Ruth. 2006. A New Greenland Ice Core Chronology for the Last Glacial Termination. *Journal of Geophysical Research* 111 (D06102): 1–16.

Reinig, F., L. Wacker, O. Jöris, C. Oppenheimer, G. Guidobaldi, D. Nievergelt, F. Adolphi, P. Cherubini, S. Engels, and J. Esper. 2021. Precise Date for the Laacher See Eruption Synchronizes the Younger Dryas. *Nature* 595 (7865): 66–69.

Renssen, H., and R. F. B. Isarin. 2001. The Two Major Warming Phases of the Last Deglaciation at ~14.7 and ~11.5 Ka Cal Bp in Europe: Climate Reconstructions and AGCM Experiments. *Global and Planetary Change* 30 (1): 117–153.

Roosevelt, A. C., J. Douglas, and L. Brown. 2002. The Migrations and Adaptations of the First Americans: Clovis and Preclovis Viewed from South America. In *The First Americans: The Pleistocene Colonization of the New World*, ed. N. G. Jablonski, 159–236. vol. 27. continuing vols. Berkeley: Memoirs of the California Academy of Sciences.

Roosevelt, A. C., M. Lima da Costa, C. L. Machado, M. Michab, N. Mercier, H. Valladas, J. Feathers, W. Barnett, M. Imazio, A. Henderson, J. Silva, B. Chernoff, D. S. Reese, J. A. Holman, N. Toth, and K. Schick. 1996. Paleoindian Cave Dwellers in the Amazon: The Peopling of the Americas. *Science* 272 (5260): 373–384.

Shaftel, H. 2022. Global Surface Temperature | Nasa Global Climate Change. In *NASA Global Climate Change*, ed. S. Callery. vol. 2022. NASA's Jet Propulsion Laboratory, California Institute of Technology.

Southey, R. 1810. *History of Brazil: Part the First*. London: Longman, Hurst, Rees, and Deme, Paternoster row.

Stute, M., M. Forster, H. Frischkorn, A. Serejo, J. F. Clark, P. Schlosser, W. S. Broecker, and G. Bonani. 1995. Cooling of Tropical Brazil (5°C) During the Last Glacial Maximum. *Science* 269 (5222): 379–383.

Szarka, L., W. W.-H. Soon, and R. G. Cionco. 2021. How the Astronomical Aspects of Climate Science Were Settled? On the Milankovitch and Bacsák Anniversaries, with Lessons for Today. *Advances in Space Research* 67 (1): 700–707.

Vizeu, R., and C. Almeida. 2006. *Relatório De Aspectos Abióticos Da Região Do Parque Estadual Monte Alegre*. Brazil: Mimeo.

Weaver, A. J., O. A. Saenko, Peter U. Clark, and J. X. Mitrovica. 2003. Meltwater Pulse 1a from Antarctica as a Trigger of the Bølling-Allerød Warm Interval. *Science* 299:1709–1713.

Zahlbruckner, A. 1909. Materiaes Para a Flora Amazonica: Vii Plantae Duckenae Austro-Guyanenses; Enumeração Das Plantas Siphonogamas Colleccionadas De 1902 a 1907 Na Guyana Brasileira. *Boletim do Museu Goeldi (Museu Paraense) de Historia Natural e Ethnographia* V (1-2): 258–261.1.

Zimmermann, K. A. 2017. Pleistocene Epoch: Facts About the Last Ice Age. In *Live Science*, ed. J. Bryner, pp. explains facts about the Pleistocene. vol. 2022. New York: Future US Inc.

Chapter 3
Phenology and Ecology of Monte Alegre State Park

Introduction

They are not the tallest hills in the chain, but relative to the riparian terrain around them, *Paituna* and *Ererê* would appear to jut into the sky from the perspective of a canoe on the river. The hill's imposing appearance would have been attractive to an ancient Stone Age culture navigating down the Amazon River. The third hill, *Serra da Maxirá*, has only one painted cave. Recently, rock art has also been found at *Caucuare*, a hill much further to the north. The tallest hill in the Monte Alegre dome, *Serra Itauajuri*, is a grassy hill with no cliffs and barely any trees.

From their stable new settlement at *Paituna*, Paleoindian pioneers could now explore inland. The next closest ridge, *Serra do Ererê*, is larger than *Paituna*; and on its western edge, Paleoindians would have discovered sturdier and more abundant silicified sandstone than that at *Painel do Pilão*. Just like on *Paituna*, most of the rock art on *Serra do Ererê* is found on the side of the hill that has resources beneficial to a Stone Age culture.

Here also, Paleoindians would have seen naturally occurring red and yellow lichen growing (often in circular pairs) on rocks and trees. These two species appear to be unique to the Amazon region (*personal communication J.P. Brown Field Museum 2Oct2009*). The vibrant colors of the lichen are believed to be a function of the symbiotic benefit of the algae to protect the fungus component of the organism from harmful spectral wavelengths (Hale 1956: 671).

Many vibrantly colored lichen grow in pristine environments, so much so that lichens have been studied as *pollution-indicators* (Jovan 2008), since they can be dramatically affected by subtle climate changes, just like coral reefs can be affected. Due to this, it stands to reason that the lichen species present here today were most likely also present in the past, perhaps to a greater degree, since I assume the earliest inhabitants came upon a more pristine environment.

C. S. Davis, *Before Sunset*, Conflict, Environment, and Social Complexity,
https://doi.org/10.1007/978-3-031-93373-8_3

Another intriguing occurrence at Monte Alegre is the frequent halo effect. Both sun and moon halos can occur naturally in most places on earth. They form in areas where moist air commonly raises high in the atmosphere to form cirrostratus clouds. At this elevation, hexagonal ice crystals form, which as a result of their shape refract light in predictably 22° or 44° circles around large bright objects in the sky (Tape and Moilanen 2006). The halos can appear rainbow-colored or grayish white depending on how cloudy the skies are. However, at Monte Alegre, the meteorological conditions that produce halos seem to occur fairly often—I have observed sun halos lasting several hours and reoccurring several days in a row. I also observed a few moon halos as well, which due to the dark backdrop of the night sky can be even more conspicuous than sun halos. Even if Paleoindians had observed halos previously at other places, they would probably have considered the frequent occurrence of them here at Monte Alegre to be enthralling.

As stated in the introductory chapter, Monte Alegre is two degrees south of the equator. Convective solar heating occurs more often here in the rainy season from December to June, with rainfall gradually increasing until the month of March (Milhomens et al. 2008: 48; Oliveira Junior et al. 1999). The warm rising air in this region is caused by the Intertropical Convergence Zone (ITCZ) which creates the highest lift of moist air in December when the air pressure is lowest, averaging a barometric pressure of 992.5 mmHg (Milhomens et al. 2008: 103). Sun or moon halos are more likely observed here during this period of time, and even more so at these hills, whose elevation and rock albedo might redirect moisture slightly higher into the atmosphere. The higher elevation also allows for a more expansive view of the sky than from under the forest canopy of the low-lying areas. The halos produce similar prismatic effects to those of a rainbow, but usually less vibrant and lacking green and blue hues. This often emphasizes for the observer the two other primary colors—red and yellow—the same two colors that appear in the lichen pairs, and also in the naturally occurring ochre and limonite clay that forms here.

To answer the question of why Paleoindian artists painted here: navigators of the Amazon River might have been drawn to the hills due to the benefits of shelter, vantage, broad-spectrum foraging, and the raw stone that they contain. Once here, they might have become inspired or motivated to artistically imitate the unusual sky and landscape features. In doing so, and through the course of time, people might have begun to make sense of those "peculiar" occurrences, associating some—such as the lichen, the halos, and the drier, durable stone of *Ererê's* western flank—to the sun or moon.

The Earth Sheds Yellow and Red: The Importance of Phenology

Geological and biological samples, some taken from animal traps between 1994 and 2006, identified plant and animal species at *PEMA* (Parque Estadual Monte Alegre = Monte Alegre State Park) over the past 20 years. Most have been endemic to the greater Amazon basin region for a long time, although state records also report that the frequency and the overall diversity of species have greatly declined in recent centuries. The geology, terrain, plants, insects, and animal species there play an important role when considering the natural resources that might have inspired the rock artists and may have provided them with animism models and prehuman paths or trails (see Milhomens et al. 2008). If there is evidence that the earliest settlers at Monte Alegre interpreted and made use of natural patterns in the landscape, then some form of belief in animism or animatism possibly provided a functional mental epistemology for their survival and success.

Natural patterns that potentially influenced human behavior are explored in the study of *phenology*. *Phenology* is the study of repeated natural patterns in relation to climate, plants, and animals. These often-overlooked natural patterns can provide human settlers (who are new to a region) with the impression that the landscape had been prepared for their arrival. Alternatively, when causal factors have long since vanished from the landscape or gone extinct (like trails gouged through the rainforest by precipitation runoff, or by megafauna, or even by leafcutter ants), animism or animalism culture might conclude that the land was previously occupied by some mythical beings.

For example, there is a myth, told among the *Craho* culture, of a hunter who went into the hole of an armadillo and found a bunch of peccaries living there, and therefore this is how hunters know where to find peccaries today (Wilbert and Simoneau 1987: 96–98). The *Craho* are one of the Gê-speaking tribes, currently scattered through northeastern and eastern central Brazil, yet their myths share a "remarkably high degree of consistency in both form and content (Freire 2001: 9)" with other Gê-speaking cultures. There are several versions of another Gê myth told by the *Craho*, *Cayapo*, *Ramkokamekra*, and *Apinaye*, where a hunter was pursuing an armadillo that fell into a hole that led to a large plain below. Young members of the hunter's tribe got rope and climbed down the hole (in some versions they climbed down a *Buriti* palm tree) and this is how people descended from the sky to populate the earth (Wilbert and Simoneau 1987: 95–111).

Near the end of the Ice Age, Paleoindians pioneered into South America as the first humans, and they might have discovered trails formed by *Neochoerus*, the 6-foot-long megafauna capybara, or caves rubbed and widened by *Megatherium*, the ground sloth that could grow up to 3 m tall (Fariña et al. 2013). Recent evidence has revealed that South American Paleoindians 13,200 years ago indeed preserved the hide of a giant ground sloth at Mylodon Cave in Chile (Pérez et al. 2020). Another research article argues that South American Paleoindians made "pendants from giant sloth osteoderms (Pansani et al. 2023)." Even the armadillo-like megafauna, *Glyptotherium*—with a weight and size comparable to a small car—may have dug underground burrows readily useable sometime later by early human pioneers (Fariña et al. 2013). Paths which when followed may have brought Paleoindian hunters upon unsuspecting prey. That is, after all, implied by the Craho myth of the armadillo mentioned above.

Another example of a phenology was revealed in the study of African elephants' movements through tropical forests in Gabón (White 1994), the Congo (Blake and Inkamba-Nkulu 2004), and Uganda (Wing and Buss 1970). The germination and dispersal of the *Sacoglottis gabonensis* seed is assisted by the fruit passing through the digestive track of elephants as they trek through the forest (Morgan 2007; White 1994), just one of many plant species dispersed by elephants acting as "mega-gardeners" (Campos-Arceiz and Blake 2011). In return, the seeds sprout plants concentrated along the same paths the elephants passed through before, with every successive pass propagating more of the fruit they like, which in turn incentivizes elephants to continuously repeat these same paths, further widening them (Blake and Inkamba-Nkulu 2004; Campos-Arceiz and Blake 2011; Vanleeuwe and Gautier-Hion 1998; Wing and Buss 1970). Additionally, the entire forest benefits from the widened paths (Vanleeuwe and Gautier-Hion 1998) acting as natural firebreaks that abate the spread of wild fires (Cardoso et al. 2020). The paths naturally produced by elephants are just one of an innumerable number of natural relationships existing in a tropical rainforest teeming with life.

Humans entering these natural landscapes, especially for the first time, would have come upon paths with which to pattern their activities, sometimes uncertain who their benefactor was. Today, it is hard to imagine what it would have felt like to be among the first human groups entering a new habitat. Would such pioneers have felt afraid, excited, alert? What would they be thinking about? Food, shelter, safety, discovery? Would they have even suspected themselves to be the first humans to enter the territory? Or would they have mistaken animal trails and other signs for evidence of prior humans, spirits, or forces by which to follow and learn from? If human ancestors all truly held traditional animism (or perhaps animatism) beliefs, it's possible that they would have interpreted natural patterns as useable information and guidance in the new land. Natural patterns may have included animal and/or weather-worn trails, hollowed-out caves, and landscape marks (like white bird droppings or lichen—see Figs. 3.1 and 3.2).

Flora

Terrain in the state park today consists predominantly of *cerrado*/savannah biomes, with reduced pockets of tropical rainforest, mangroves, swamps, and marshes—all types showing various degrees of human alteration and secondary growth (Milhomens et al. 2008: 51; Pastana 1999). However, it has been argued (Roosevelt 2000: 462–464) that any significant changes to the environment only began during

Fig. 3.1 Naturally occurring lichen that grows often in red and yellow pairs, shown here on a rock on the slope of *Serre do Ererê* near the rock art of *Serra da Lua* (above) and on a tree in the valley at the base of the hill, with some purple hues (below)

late prehistoric times and rapidly increased during much more recent human deforestation activity.

Rainforests, mangrove swamps, and marshland biomes experience seasonal flooding, and they are the least altered by humans. In the state park, these biomes are predominantly encroached by the communities of *Paituna* and *Santana* at the

Fig. 3.2 Naturally occurring lichen growing in red and yellow pairs, shown here on trees in the woods near *Serra da Paituna*

southeastern flank of *Serra da Paituna* (Milhomens et al. 2008: 51) and *Lages* to the southwest flank. Other vegetation in the region includes pioneer scrublands and wooded grasslands—both indicators of heavily disturbed terrain in the process of regeneration. This terrain has been intentionally introduced by human practices of slash-and-burn and shifting cultivation to produce grazing meadows for livestock.

Tropical Rainforest and Wetlands

The local tropical rainforest biome includes several common tree and shrub species including flowering trees from the *Caraipa* genus (Milhomens et al. 2008: 131). *Caraipa spuria*, a spruce-like evergreen tree, grows in seasonally flooded low-lying blackwater swamps (see Fig. 3.3), and there is also a shrub-like bush in the *Caraipa* genus that similarly grows in seasonally flooded white-sand savannas near streams (Kubitzki 1978). Crabwood trees, like *Andiroba* from the mahogany family, and other trees from the genus *Carapa* are used for their mosquito-repellent oil. *Andiroba* candles are made and sold locally in the city of Monte Alegre.

Fig. 3.3 Riverboating through the marsh from the Lages community to the city of Monte Alegre during the flood season in May 2011

Other plants common to the local tropical forest and mangrove habitats include the Sandbox tree[1] (*Hura crepitans*); the resinous *breu-branco* tree (*Protium spruceanum*); various types of leguminous trees (from the genus *Vatairea*); the silk floss tree (*Ceiba speciosa*), along with the related *baobab* and *kapok* trees; the *ucuuba* tree (*Virola sebifera*)—with its hallucinogenic bark;[2] and many varieties of palm trees. The more prominent palm trees include: the *jauari* palm[3] (*Astrocaryum jauari*), the *açai* palm (*Euterpe oleracea*), the *inaja* palm (*Attalea maripa*[4]), the *buriti* palm[5] (*Mauritia flexuosa*), the *piririma* palm tree (*Syagrus cocoides*), the *urucuri* palm (*Attalea phalerata*), and the *curuá* palm (*Attalea monosperma*) (Milhomens et al. 2008: 131). Extracts from the fruit of the *jauari*, *buriti,* and the *açai* palm trees all have found very successful commercial use.

Disturbance to the natural forest from strong riparian and alluvial forces (as well as from human activity) can put into motion a succession of vegetative species—of which the early stages often include the advancement of pioneering scrubland. This succession appears to be prevalent within the research area. The most common species in this scrubland biome include the leguminous shrub (*Machaerium lunatum*), the liana vine (*Dalbergia monetária*), and the helophytic aquatic plant (*Montrichardia arborescens*) (Milhomens et al. 2008: 134, 135).

Woody Grassland Savannah

This biome contains various species of the *Paspalum* genus in sometimes dense mats reaching more than a meter in height. During seasonal floods that turn the terrain into swamp, the soil can become clayey-loam, and the dense mats can become floating islands. These pockets of vegetation contribute to the deposition and fixation of water-transported sediments, facilitating the establishment of plants with more developed root systems. The most common species cited in this biome includes: barnyard grasses (*Echinochloa polystachya*), horsetail grass (*Paspalum repens*), crowngrass (*Paspalum fasciculatum*), relatives of switchgrass (*Panicum zizanioides*), West Indian marsh grass (*Hymenachne amplexicaulis*), a species of rice (*Oryza*), several species of water hyacinth (*Eichornia*), several species of water

[1] Also called the dynamite tree (*Açacu* in Portuguese) because its seeds explode from the fruit when ripe—some report the explosion launches seeds at 70 m/s at a distance that can reach 100 m. Fishermen use the sap to poison fish and Caribs used it in their poisonous arrows (Wikipedia).

[2] Ucuuba bark is said to have the hallucinogenic dimethyltryptamine (DMT), and it has been used in traditional medicines and burned to ward off evil spirits (Wikipedia).

[3] Which produces the commercially edible and nutritious "heart of palm."

[4] Also known under an older designation as *Maximiliana maripa*, or maripa palm.

[5] Also called moriche or boriche, it tends to grow very tall in natural water springs, alerting a thirsty traveler to its presence.

lily (*Nymphaea*), pickerel weeds (*Pontederia*), the water cabbage (*Pistia stratiotes*), yellow flowering plants related to the pea plant (*Neptunia*), and one of the largest lily pads in the world (*Victoria amazônica*).

Cerrado/Savannah

Cerrado and savannah make up much of the research area terrain. These terms are sometimes used synonymously; however, *cerrado* is a very diverse, dense-growth vegetative savannah. In this biome, most plants grow to a small or medium size, not necessarily tall canopy trees, but ground cover is almost completely continuous—hence the translation of the word "*cerrado*" means "closed." Savannah indicates the occurrence of an herbaceous stratum dominated by grasses and a tree/shrub stratum that is either continuous or discontinuous.

There are pockets of true savannah in the state park that suggest the biome underwent rigorous selective pressures. In the hills, the eroded, loose, sandy soil provides ideal conditions for *cerrado* vegetation dominated by sedges and true grasses (*Poaceae* and *Cyperaceae* genera). At lower elevation, shrubs dominate the *cerrado* vegetation. Still in other special low-lying areas—especially along forest margins—a type of *cerrado* forms from well-developed trees whose branches intertwine with bushes to create dense thickets. Near rocky outcroppings and in the crevices of large boulders, and also in caves, microenvironment conditions further specialize the habitable plants (Milhomens et al. 2008: 135).

At the boundary between the forest and the highlands, irregularly eroded soils dictate *cerrado* vegetation, reducing hydrophilic plants into smaller clearings interspersed with pastureland. Plant species registered in this zone include: *Krameria tomentosa, Heisteria flexuosa, Ixora martensii, Vernonia fruticulosa, Anacampia coriacea, Tibouchina spruceana, Trachypogon vestitus, Protium heptaphyllum,* and *Dioclea glabra*. Trees are sparse in this terrain, with the predominant species being the *Sacuri* palm tree (*Attalea microcarpa*). The dominant grasses are *Trachypogon spicatus* and *Fimbristylis capillaris* (Milhomens et al. 2008: 135).

At higher elevation, savannah is typically characterized by small irregularly spaced groups of diverse trees and bushes. The most common species in this savannah include: *Casearia grandiflora, Alibertia edulis, Qualea grandiflora, Eugenia flavescens, Eugenia tapacumensis, Vismia baccifera, Memora allamandiflora, Memora magnífica, Byrsonima crassifolia, Lafoensia densiflora, Pouteria lateriflora, Sacoglottis guianensis, Davilla pedicellaris, Ocotea fasciculata, Plumeria fallax, Limnosipanea schomburgkii, Paepalanthus polytrichoides, Cuphea micrantha, Polycarpaea corymbosa,* and *Sclerolobium paniculatum*.

Some cacti have also been registered in this zone, including plants from the *Cereus* genus. In the heights of the hills, the red-hot poker vine (*Norantea guianensis*) grows, as well as several ferns (Milhomens et al. 2008: 136).

Disturbed Forests (mata *and* capoeirão)

Disturbed forests are forests that have been altered by a human presence, leading to the propagation of secondary vegetation. The most persistent disturbance to the forest comes not directly from humans, but indirectly from the grazing of peccaries raised by humans.

Local ranchers and farmers also induce brush fires in areas to keep vegetation low enough for grazing. The overall impact leaves a range of forest classified as low to moderate intensity secondary forest. These altered forests nucleate around rivers, roads, and settlements—gradually decreasing at a gradient extending outward, so this biome was probably much smaller (or nonexistent) during the Paleoindian period. The most concentrated disturbed vegetation is locally referred to as "mata" (woods) or "capoeirão" (coppice) (Milhomens et al. 2008: 137).

Trees in this biome can grow 20 to 30 m in height. The most frequent species registered in the "*mata*" (woods) include: *Eschweilera ovata, Bellucia grossularioides, Tapirira guianensis, Sclerolobium paniculatum, Rhabdodendron amazonicum, Copaifera martii, Sacoglottis guianensis, Myrcia sphaerocarpa*, and the *Sacuri* palm tree (*Attalea microcarpa*). In the subforest the most frequent species are: *Monotagma laxum, Abuta grandifolia, Psychotria barbiflora*, and *Bactris cuspidata* (Milhomens et al. 2008: 138).

The "capoeirão" (coppice) refers to the transition boundary from "mata" to "cerrado" vegetation, where dense bushes and trees only reach about 5 m in height. The main species here are: *Ouratea castaneifolia, Vismia baccifera, Anacardium occidentale, Copaifera martii, Rhabdodendron amazonicum, Himatanthus sucuuba, Hymenaea parvifolia, Qualea grandiflora, Vismia baccifera, Byrsonima crassifólia*, plants from the *Tabebuia, Myrcia* and *Eugenia* genera, and plants from the *Dilleniaceae* and *Chrysobalanaceae families* (Milhomens et al. 2008: 138).

Cave Vegetation

Some of the caves in the hills provide special microclimates that support unique flora. Plants commonly found in these habitats include *Begonia guianensis, Adiantum tenuissimum, Ichnanthus hoffmannseggii, Anthurium fragrantissimum*, and a species of *Paepalanthus. Pedra Pintada* cave has scant vegetation. Species that have been reported to grow there include a species of *Begonia* plant, *Aciotis aequatorialis*, the cabbage palm fern[6] (*Phlebodium aureum*), flowering plants from the *Araceae* family, and mosses. The cave walls of *Itatupaoc*a have a thin organic layer that supports a variety of species, including: *Begonia prieurii, Aciotis aequatorialis,*

[6] Known as *Samambaia* in Brazil—this plant has been commercially found to treat psoriasis and other skin conditions like Vitiligo, Alzheimers, dementia, Multiple Sclerosis, autoimmune disorders, and upper respiratory problems.

Koellikeria erinoides, Tabernaemontana angulata, Selaginella tenuissima, and a species of *Paepalanthus*. In the interior of the cave, where light hits the surface, a layer of bryophytes (mosses and liverworts) from the genus *Ricardia* tend to grow. Also, seeds of the tree species *Sacoglottis guianensis* and the lacquer-producing *Visima bacífera* tend to grow inside and sprout (Milhomens et al. 2008: 138).

Other caves show significant growth of creeping and climbing wall plants, such as *Ichnanthus leptophyllus* and members of the *Hymenophyllaceae* family of plants. Some of the cracks and crevices in caves can develop a thick tangle of roots and branches descending down from plants of the *Moraceae* family growing on the hills above. Many caves in the region also tend to accumulate seeds from the cabbage tree[7] (*Andira inermis*), *Sacoglottis guianensis*, the *curuá* plant (*Ananas lucidus*) whose fibers have been used by indigenous tribes throughout the tropical Americas, the *jatá* palm (*Syagrus cocoides*), the *sacuri* palm (*Scheelea martiana*), and the *tucumã* palm (*Astrocaryum tucumã*) (Milhomens et al. 2008: 138). Most of these seeds are transported into the caves by rodents, bats, and other small animals.

Lichen

Lichen grow abundantly on many trees and exposed rocks throughout the hills (see Fig. 3.4). Specific information on their identity had not been previously provided in research, except for the greenish-white type, *Cladonia coccifera*, identified by Zahlbruckner (1909: 261). I took samples in 2009 and had them identified by the Botany department at the Field Museum of Natural History in Chicago. The red lichen was identified as *Laurera purpurina* and the yellow lichen is *Chrysothrix xanthina*. While they may exist separate in other locations in the world, I was told my samples were specific to the Amazon.

Fauna

Insects

The abundant foliage in the region offers innumerable microhabitats for fauna. These habitats offer a home for an abundance of insect orders (see Fig. 3.5) such as: mosquitos (*Culicidae*), flies (*Diptera*); wasps, bees, and ants *(Hymenoptera*); beetles (*Coleoptera*); and moths (*Lepidoptera*).[8] Marshland microhabitats add niches

---

[7] Locally this tree is called morceguiera, because bats are known to eat its fruit. Their digestion is responsible for some seed accumulations. The bark of the tree is also known to have narcotic, laxative, and vermifuge properties.

[8] Ancient Greek root word meanings provide more accurate detail: *diptera* (two wings), *hymenoptera* (membrane wings), *coleoptera* (sheathed wings), and *lepidoptera* (scaly wings).

Fig. 3.4 A tree covered with red, yellow, and greenish-white lichen in the deep forest of Serra da Paituna

Fig. 3.5 (**a**) Moth resting on clothes on a clothesline. (**b**) *Automeris Io* caterpillar crawling on the side of the dirt road. (**c**) dragonfly perched at the end of a tree twig

for various millipedes (*Diplopoda*), and sandy soils provide a niche for predatory tiger beetles (*Cicindelidae*).

Mosquitos[9] (of the suborder *Nernatocera*), dragonflies (*Odonata*), fruit flies (*Drosophila*), and horse flies (*Tabanidae*) are the most common insects seen

[9] Locally called "carapanã" in Monte Alegre, these are almost certainly a local species different from the regional "pernilongo" mosquitoes.

(Milhomens et al. 2008: 51). Arachnids—including mites and ticks (*Acari*), spiders (*Araneae*), harvestmen (*Opilones*), and whip spiders (*Amblypygi*)—inhabit the many caves and grottoes of the hills. Fourteen families from the *Opilones* order also occupy the leaf-litter and fallen tree trunks in the closed forests as well as in the floodplains and along lake margins (Milhomens et al. 2008: 51).

Fish

Commercial fish have been registered in the shoals and bays of the Amazon River between Santarém and Monte Alegre in upwards of 128 species (Ferreira et al. 1998: 211). The most common fish captured for consumption in the waters surrounding the state park include the *acará* (a cichlid fish), *acari-bodo* (*Pterygoplichthys pardalis*—a suckermouth pleco fish), *aracú* (*Schizodon fasciatus*—a fish of the anostomids), *aruanã* (or *arowana*—a fish from the Osteoglossidae family of bonytongues), *charuto* (a characid fish from the Hemiodontidae family), *cujuba* (a fish from the Doradidae family), *curimatã* (*Prochilodus nigricans*—a flannel-mouthed characid fish), *dourada* (*Brachyplatystoma flavicans*—a type of catfish), *jaraqui* (a type of characid fish from the genus *Semaprochilodus*), *jiju* (a fish from the Erythrinidae family), *mapará* (*Hypophthalmus edentates*—a long-whiskered catfish), *pacú* (*Metynnis hypsauchen*—the "silver dollar" characid), *pescada* (a croaker fish from the Sciaenidae family), piaba (*Astyanax jacobinae*—a characid fish), *piranha* (the popular meat-eating fish from the *Serrasalmus* genus), *pirapitinga* (*Piaractus brachypomus*—a close relative to piranhas), *pirarucu* (*Arapaima gigas*—the largest freshwater fish in the world), *surubím* (*Pseudoplatystoma fasciatum*—a long-whiskered catfish), *tamoatá* (*Hoplosternum littorale*—a type of catfish), *traíra* (*Hoplias malabaricus*—also referred to as wolf fish or tiger fish), and the *tucunaré* (from the genus *Cichla*—a group collectively called "peacock bass" due to a false "yellow eye" on their tail). All of the above-mentioned fish are freshwater fish and most have the ability to breathe air during periods of shallow water (Milhomens et al. 2008: 181).

Reptiles/Amphibians

Numerous herpetofauna also inhabit the wetlands and closed forests. A total of over 150 species of snakes (Martins and Oliveira 1998) and 89 species of lizards (Avila-Pires 1995) exist in the region. Additionally, all the major reptilian orders are present in the waters around Monte Alegre, including turtles (*Chelonia*), snakes (*Serpentes*), lizards (*Lacertilla*), and crocodiles or caiman (*Crocodylia*). There are at least 4 species of crocodile (or caiman)—all carnivorous—and about 15 species of turtles—mostly vegetarian (see Fig. 3.6a). Only one study had been conducted so far (Mesquita et al. 2006) to reveal at least 22 amphibian species and 20 reptile

Fig. 3.6 (**a**) Left photo of the Arau turtle (*Podocnemis expansa*) captured near Lake Miritiepe. (**b**) Center photo of the turnip-tailed gecko, taken at *Painel do Pilão* excavation. (**c**) Right photo of unidentified frog species on the side of a house in the *Lages* community

species living within the limits of the state park. Common lizards there include the Amazon racerunner (*Ameiva ameiva*) and the cryptic racerunner (*Cnemidophorus cryptus*).

Three lizard species are particularly common in the hills: the lava lizard[10] (*Tropidurus hispidus*) that frequently peaks out from rocky outcroppings, the striated whiptail lizard (*Kentropyx striata*), and the turnip-tailed gecko (*Thecadactylus rapicuada*—see Fig. 3.6b). Also, the venomous tropical rattlesnake[11] (*Crotalus durissus*) and five amphibian species—including the Myers' thin-toed ditch frog (*Leptodactylus myersi*)—live in the hills (see Fig. 3.6c). The tropical rainforest ecology near *Ilha Grande* additionally supports the phantasmal poison dart frog (*Epipedobates hahneli*), a separate species of frog from the *Adenomera* genus, and two seven-striped blind snakes (*Leptotyphlops septemstriatus*) were captured there as well (Milhomens et al. 2008: 141–142).

Birds

The registry of birds in the region is over 261 species from 59 families (Milhomens et al. 2008: 145). Some of the more prominent birds are the numerous insectivorous perching birds (*Passeriformes*) such as the flycatcher (*Tyrannidae*) and the antbird (*Thamnophilidae*). There are also a number of fruit-eating birds, including the parrot, parakeet (*Aratinga pintoi*), pigeon, and other *Passeriformes*. *Anambé* birds

[10] An iguanian type of lizard, which Brazilians generally call "calango."

[11] Called "cascavel" in Brazil—it is actually a pit viper. I heard this snake rattle very close to me while I was exploring a dark cave that the locals called *Caverna da Onça* on top of *Serra da Paituna* near *Labirinto* cave.

Fig. 3.7 A flock of vultures (*Urubu—Coragyps atratus*) assemble on posts near a dead cow in the *Lages* community

(umbrella bird[12]), mockingbirds (*Mimidae*), and thrush birds (*Turdidae*) often lurk in the forest canopy, while *Cariamidae* (seriema) are the flightless terrestrial birds. Representatives of the *Cuculidae* family (cuckoos) inhabit both the trees and the ground. Heron (*Ardeidae*) and kingfishers (*Alcedinidae*) are the most common aquatic birds; other aquatic birds include several members of the *Anatidae* (ducks, geese, swans) and *Threskiornithidae* (ibises and spoonbills) families (Dantas 2006; Milhomens et al. 2008: 52).

Birds-of-prey include several representatives from the *Accipitridae* family, which include hawks, kites, eagles, and some vultures (see Fig. 3.7), and the *Falconidae* family (falcons and kestrels). Three species of owls had been recorded in the region: the small-sized tropical screech owl (*Megascops choliba*), with pronounced ear tufts and whose courtship singing lasts from August to September; the medium-sized mottled owl[13] (*Strix virgata*), which has no ear tufts, appears mostly white and inhabits the highest elevations; and the large-sized spectacled owl[14] (*Pulsatrix perspicillata*), which has angled white eyebrows, no ear tufts, and is almost exclusively solitary and nocturnal (Milhomens et al. 2008: 278).

[12] Of the genus *Cephalopterus*—the species is uncertain but probably *ornatus*.

[13] Called "coruja-do-mato" in Brazil. I am almost certain this was the owl I saw perched in the cave—Abrigo do Coruja—in Serra da Paituna. On three occasions when I entered the cave, the owl fled before I could capture a picture of it with my camera.

[14] Called "murucututu" in Brazil.

Omnivorous birds include blackbirds, the Cayenne Jay (*Cyanocorax cayanus*), and some woodpeckers (Milhomens et al. 2008: 145).

Mammals

Mammals in the past were certainly more diverse and numerous than mammals inhabiting the region today. A recent registry (Souza et al. 2006) found that mammal orders have been decreasing in the area—from eleven to seven—due to hunting and destruction of the natural habitat. Local inhabitants of the region had reported the existence of a wide range of species that could not be officially confirmed during recent investigations. These animals, in strong decline, include: the Amazon River pink dolphin[15] (*Inia geoffrensis*), the *tucuxi* dolphin (*Sotalia fluviatilis*), the Amazon river manatee[16] (*Trichechus inunguis*), a tapir (*Perissodactyla*), two species of otter (*Lontra longicaudis* and *Pteronura brasiliensis*),[17] the capybara (*Hydrochaeris hydrochaeris*), and the Brazilian rabbit[18] (Sylvilagus brasiliensis) (Milhomens et al. 2008: 52).

The species that are still thriving in the area, based on the frequency of recorded sightings, include the: collared anteater (*Tamandua tetradactyla*), golden-handed tamarin[19] (*Saguinus midas*—Fig. 3.8), crab-eating fox (*Cerdocyon thous*), white-lipped peccary (*Tayassu pecari*—Fig. 3.9), and both the red and brown brocket deer (*Mazama americana* and *Mazama gouazoubira,* respectively). Rodents include the Brazilian squirrel (*Sciurus aestuans*), the paca/agouti (*Agouti paca*) and the cutia/cuchilla (*Dasyprocta leporina*) (Milhomens et al. 2008: 52, 152). Some resident hunters also had captured some coati (*Nasua nasua*) and kept them as pets. The tayra[20] (*Eira barbara*) is also listed as common. And finally, the official registry also confirmed the presence—albeit rare—of several felines: the jaguar[21] (*Panthera onca*), the cougar[22] (*Puma concolor*), the ocelot[23] (Leopardus pardalis), the

[15] Called "boto" in Brazil. In the order of Cetacea, which includes whales and porpoises as well—both of which were rumored to swim this far up the Amazon too.

[16] Called "peixe-boi" in Brazil. In the order of *Serenia*, it is renowned regionally for its succulence, and therefore has been overhunted.

[17] Called the "lontra" and the "ariranha," respectively, in Brazil.

[18] Called the "tapeti," it is crepuscular and the only lagomorph in the Amazon forest.

[19] Called "sagüi" in Brazil.

[20] Called the "irara" locally.

[21] Generally referred to as "onça" in Brazil, or "onça-pintada."

[22] Called "suçuarana" in Brazil.

[23] Called "jaguatirica" in Brazil.

Fig. 3.8 Golden-handed tamarin (*Saguinus midas*) being coy while photographed in *Serra da Paituna*

Fig. 3.9 Peccary. Mother and offspring (domesticated) freely roaming through Lages near *Serra da Paituna* during the rainy season

Fig. 3.10 Tufted Brown Capuchin Monkey (*Cebus apella*) spying on the excavation of *Gruta 15 de Março* on *Serra da Paituna* hill

margay[24] (*Leopardus wiedii*), and the jaguarundi[25] (*Herpailurus yagouaroundi*) (Milhomens et al. 2008: 152).

Rare encounters with several primate species were confirmed in the region as well. In addition to the thriving tamarin species mentioned above, other primates include: the white-faced saki monkey[26] (*Pithecia pithecia*), the red-backed bearded saki monkey[27] (*Chiropotes sagulatus*), the common squirrel monkey[28] (*Saimiri sciureus*), and the tufted capuchin monkey[29] (*Cebus apella*—Fig. 3.10) (Milhomens et al. 2008: 150, 151). The howler monkey was also mentioned as rare, but my experience with them was quite frequent, as I recount at the end of this chapter.

A number of nocturnal animals have also been confirmed in the region, including several opossum species: the delicate slender opossum[30] (*Marmosops parvidens*), the brown four-eyed opossum[31] (*Metachirus nudicaudatus*), Linnaeus's mouse opossum (*Marmosa murina*), the northern red-sided opossum (*Monodelphis brevicaudata*), Cuvier's spiny rat (*Proechimys cuvieri*), and possible occurrences of the water opossum (*Chironectes minimus*), and the bare-tailed woolly opossum (*Caluromys philander*) (Nunes 2001). Also nocturnal, the silky anteater (*Cyclopes didactylus*), the Brazilian tree porcupine (*Coendou prehensilis*) with its prehensile

[24] Called "maracajá-peludo" in Brazil.

[25] The name "jaguarundi" comes directly from the Tupi Indian name for the cat "Yawaum'di." Another Brazilian name for it is the "gato-mourisco."

[26] Called "parauacú" or "cara-branca" in Brazil.

[27] Called "cuxiú" in Brazil.

[28] Called "macaco de cheiro" or "macaco amarelinho" in Brazil.

[29] Called "macaco-prego" in Brazil.

[30] Called the "mucura" in Brazil, although there is more than one species.

[31] The species translation is not certain. The source lists (*Didelphimorphia, cuíca*), but later refers to a "mucura-quatro-olhos"—which translates to "4-eyed opossum."

Fig. 3.11 A vampire bat hanging upside down on a barbed-wire fence in the *Lages* community during the morning

tail, the 15-kg greater long-nosed armadillo (*Dasypus kappleri*), and the southern naked-tailed armadillo (*Cabassous unicinctus*) have all been confirmed to be living in the state park. The kinkajou[32] (*Potos flavus*) has also been discovered to roam the area, sometimes mistakenly identified as the night monkey (*Aotus*). And lastly, confirmed to be living in the area is the (sometimes nocturnal) brown-throated sloth[33] (*Bradypus variegatus*) (Milhomens et al. 2008: 150–152). Several nocturnal bat species also inhabit the state park, particularly in caves, grottoes, and shelters. A total of five different species of bats have been identified so far: the Jamaican fruit bat (*Artibeus jamaicensis*), the Seba's short-tailed bat (*Carollia perspicillata*), the common vampire bat (*Desmodus rotundus*—see Fig. 3.11), the pygmy round-eared bat[34] (*Lophostoma brasiliense*), and the Parnell's mustached bat (*Pteronotus parnelli*) (Milhomens et al. 2008: 151). While excavating in *Gruta 15 de Março*, I was able to observe a community of bats for nearly a month. The floor of the cave—as well as the floors of several other caves—had a layer of soil infused with their bat guano, which gave the cave a gunpowder odor.

Additional Animals Encountered

During my time exploring and excavating in the hills, I also encountered several other species beyond those mentioned above. A group of tufted capuchin monkeys (*Cebus apella*) spied and peeked in on our excavation at *Gruta 15 de Março*. The

[32] Called "jupará" in Brazil.

[33] They sleep ~18 h a day and come to the ground from the trees once a week. The female mating call is said to sound like a woman screaming "ay ay" (Wikipedia).

[34] Reported as *Tonatia brasiliensis* but the scientific name has since been changed.

weeper capuchin monkey[35] (*Cebus olivaceus*) is claimed not to exist in the park (Milhomens et al. 2008: 151) but we suspect we saw a few of these watching us at a distance while excavating there as well.

However, due to the distance and partial canopy concealment, we might have mistaken them for tufted capuchins. The most memorable animal I encountered in *Gruta 15 de Março* was a black frog with red marks on either side of its belly; it reached the ceiling of the cave to jump at, and snag in its mouth, a bat that was clinging upside down there. I have not been able to identify this unknown species of frog, but I did also encounter another one in a different, very small and very moist, cave near the northern end of *Serra da Paituna*.

On several occasions while exploring, I came upon a roaming band of *howler monkeys* (*Alouatta seniculus*) near the swamp by the Lages community. I also encountered a band of them in the disturbed forest, and a possibly different band in the closed forest at the base of the hills. Although they are reported to exist in the park, the report claims they are scarce (Milhomens et al. 2008: 151). I found this to be grossly understated, as the *howlers* were the most represented during my stay in Lages. Their roars were heard at dawn almost every day and often also in the few hours after dusk in the months of April and May—perhaps their mating season. While walking near the edge of the swamp, we came upon a copse of *jauari* palm trees with over 20 *howler monkeys*.[36] They fled from us immediately, scampering down the tree like a vertical stampede.

I also encountered some bush dogs[37] (*Speothos venaticus*) on more than one occasion in the disturbed forest on the western flank of *Serra da Paituna*, which at the time I assumed were just small hogs. I remember that these animals stayed within proximity to the ranches in that area, leaving me with the impression that they were possibly domesticated. However, after learning of their existence in the park and what they look like, it is probable that these dogs were preying on the young, domesticated hogs of the ranches.

We also came upon a dead armadillo[38] (*Dasypus novemcinctus*) on the road at the base of *Serra da Lua* in *Ererê*, and while traveling to our excavation site above *Serra da Lua* one morning, we saw an adult and an adolescent red-faced spider monkey[39] (*Ateles paniscus*) slinking through a nearby tree. These primates were also claimed not to exist in the park (Milhomens et al. 2008: 151), but we are certain that is what we saw. Also, while exploring atop *Serra do Ererê*, I continuously encountered foxes; so much so that I labeled one region *Pico da Raposa* (peak of the fox), and a particular fox burrow that I found at the base of a narrow outcrop of rocks about 2 m tall inspired me to label it "fox tower." Each fox sighting was of just a solitary fox,

---

[35] Called "caiarara" in Brazil.

[36] Called "guariba" in Brazil. Their howls sound more like roars and are one of the loudest sounds produced by any animal.

[37] Called "cachorro-do-mato" in the region.

[38] Called "tatu" or "tatu-galinha" in Brazil.

[39] Called "coamba" locally.

but each encounter appeared to be different foxes that ranged in color from gray, to dark brown, to black. These could all have been from the same species (*Cerdocyon thous*) reported to exist in the region, however.

While hiking atop *Serra da Maxirá* we came across the two-toed sloth (*Choloepus didactylus*) perched low on a tree trunk (Fig. 3.12). Its docile nature and its appearance of a smile lifted our weary spirits at the time. We took a few pictures of it, and it never seemed disturbed or bothered by our presence.

Monte Alegre has a rich and diverse ecosystem with animal and plant resources that could sustain human populations at any time in history. Some species of plants, animals, fungus, and *lichen* would inevitably become more or less important than others, and so this is a good place to start when considering which subjects of nature the ancient artists might have chosen to depict. Should their art be just as diverse? Should we expect them to have depicted animals that are scarce, plants that are more vital to human sustenance, or lichen that are rare and peculiar? To gain a better understanding of this connection to the ecosystem, we must consider patterns in nature and their perceived animism.

Fig. 3.12 A sloth atop Serra da Maxirá

References

Avila-Pires, T. C. 1995. Lizards of Brazilian Amazonia (Reptilia: Squamata). *Zoologische verhandelingen* 299 (1): 1–706.

Blake, S., and C. Inkamba-Nkulu. 2004. Fruit, Minerals, and Forest Elephant Trails: Do All Roads Lead to Rome? *Biotropica* 36 (3): 392–401.

Campos-Arceiz, A., and S. Blake. 2011. Megagardeners of the Forest–the Role of Elephants in Seed Dispersal. *Acta Oecologica* 37 (6): 542–553.

Cardoso, A. W., Y. Malhi, I. Oliveras, D. Lehmann, J. E. Ndong, E. Dimoto, E. Bush, K. Jeffery, N. Labriere, and S. L. Lewis. 2020. The Role of Forest Elephants in Shaping Tropical Forest–Savanna Coexistence. *Ecosystems* 23 (3): 602–616.

Dantas, S. M. 2006. *Avifauna Do Parque Estadual Monte Alegre*. IPHAN: Relatório.

Fariña, R. A., S. F. Vizcaíno, and G. De Iuliis. 2013. *Megafauna: Giant Beasts of Pleistocene South America*. Bloomington, Indiana: Indiana University Press.

Ferreira, E. J. G., J. A. S. Zuanon, and G. M. Santos. 1998. *Peixes Comerciais Do Médio Amazonas Região De Santarém—Pa*. Brasília: IBAMA/MMA.

Freire, J. R. 2001. Da "Fala Boa" Ao Português Na Amazônia Brasileira. *Amerindia* 8:1–83.

Hale, M. E. 1956. Ultraviolet Absorption Spectra of Lichen Depsides and Depsidones. *Science* 123:671.

Jovan, S. 2008. *Lichen Bioindication of Biodiversity, Air Quality, and Climate: Baseline Results from Monitoring in Washington, Oregon, and California*, ed. U. S. D. o. Agriculture. (USDA), Pacific Northwest Research Station.

Kubitzki, K. 1978. The Botany of Guyana Highland: Part X. Caraipa and Mahurea (Bonnetiaceae). *Memoirs of the New York Botanical Gardens* 29:82–131.

Martins, M., and E. M. Oliveira. 1998. Natural History of Snakes in Forests in the Manaus Region, Central Amazonia, Brazil. *Herpetological Natural History* 6 (20): 78–150.

Mesquita, D. O., G. C. Costa, and G. R. Colli. 2006. Ecology of an Amazonian Savanna Lizard Assemblage in Monte Alegre, Pará State, BrazilBrazil. *South American Journal Of Herpetology* 1 (1): 61–71.

Milhomens, A., D. S. Nascimento, L. Salvo, W. Rodrigues, F. Ferreira, A. Mota, R. Gangana, E. R. de Carvalho, and R. d. C. M. Moreira 2008. *Plano De Manejo Do Parque Estadual Monte Alegre (Pema)*, ed. P.-P. d. D. d. E. n. A. Legal, 303. Belem: Ministerio do Meio Ambiente.

Morgan, B. J. 2007. Sacoglottis Gabonensis – a Keystone Fruit for Forest Elephants in the Re'Serve De Faune Du Petit Loango, Gabon. *African Journal of Ecology* 47:154–163.

Nunes, A. C. 2001. *Gradientes Estruturais Dos Hábitats Em Savanas Amazônicas: Implicações Sobre a Distribuição E Ocorrência Das Espécies De Pequenos Mamíferos Terrestres (Rodentia, Didelphimorphia)*. Doctoral Dissertation, Universidade Federal Do Rio De Janeiro, Rio De Janeiro, Brazil.

Oliveira Junior, R. C. d., T. E. Rodrigues, P. L. Dos Santos and M. A. Valente 1999. *Zoneamento Agroecológico Do Município De Monte Alegre, Estado Do Pará*, ed. CPATU, 87. Documentos 9 ed. Série Documentos (CPATU). vol. INFOTECA. EMBRAPA Amazônia Oriental, Belém, Brazil.

Pansani, T. R., B. Pobiner, P. Gueriau, M. Thoury, P. Tafforeau, E. Baranger, Á. V. Vialou, D. Vialou, C. McSparron, M. C. de Castro, M. A. T. Dantas, L. Bertrand, and M. L. A. F. Pacheco. 2023. Evidence of Artefacts Made of Giant Sloth Bones in Central Brazil around the Last Glacial Maximum. *Proceedings of the Royal Society B: Biological Sciences* 290 (2002): 20230316.

Pastana, J. M. d. N. 1999. Diagnóstico Do Potencial Ecoturístico Município De Monte Alegre. In *Programa de Integração Mineral em Municípios da Amazonia (PRIMAZ)*. Ministério de Minas e Energia, Belém, Brazil.

Pérez, M., N. Toledo, F. Mari, I. Echeverría, E. P. Tonni, and M. J. Toledo. 2020. Radiocarbon Dates of Fossil Record Assigned to Mylodontids (Xenarthra - Folivora) Found in Cueva Del Milodón, Chile. *Quaternary Science Reviews*.

Roosevelt, A. C. 2000. The Lower Amazon: A Dynamic Human Habitat. In *Imperfect Balance: Landscape Transformations in the Precolumbian Americas*, ed. D. L. Lentz, 455–492. New York: Columbia University Press.

Souza, B. A., C. A. L. Carvalho, and R. M. O. Alves. 2006. Flight Activity of Melipona Asilvai Moure (Hymenoptera: Apidae). *Brazilian Journal of Biology* 66 (2b): 731–737.

Tape, W., and J. Moilanen. 2006. *Atmospheric Halos and the Search for Angle X*. Washington: American Geophysical Union.

Vanleeuwe, H., and A. Gautier-Hion. 1998. Forest Elephant Paths and Movements at the Odzala National Park, Congo: The Role of Clearings and Marantaceae Forests. *African Journal of Ecology* 36 (2): 174–182.

White, L. J. 1994. Sacoglottis Gabonensis Fruiting and the Seasonal Movements of Elephants in the Lopé Reserve, Gabon. *Journal of Tropical Ecology* 10 (1): 121–125.

Wilbert, J., and K. Simoneau. 1987. Folk Literature of the Gê Indians. In *Ucla Latin American Studies 44*. Los Angeles: UCLA Latin American Center Publications.

Wing, L. and I. Buss 1970. Elephants and Forests Wildlife Monographs No. 19.

Zahlbruckner, A. 1909. Materiaes Para a Flora Amazonica: Vii Plantae Duckenae Austro-Guyanenses; Enumeração Das Plantas Siphonogamas Colleccionadas De 1902 a 1907 Na Guyana Brasileira. *Boletim do Museu Goeldi (Museu Paraense) de Historia Natural e Ethnographia* V (1-2): 258–261.

Chapter 4
Historical Accounts of the Rock Art

Alfred Russel Wallace

The earliest written mention of rock art at Monte Alegre was by the English naturalist Alfred Wallace in 1849, who, in search of cave paintings, encountered *Pedra do Pilão* and possibly *Painel do Pilão* (1979), and he was later brought to *Serra da Lua* where he sketched some of the images (Wallace 1889: 103–105). Of *Pedra do Pilão* Wallace wrote,

> We found ourselves on the platform below the columnar mass, which rises perpendicularly thirty or forty feet, and then hangs over at the top all round in a most curious and fearful manner. Its origin is very plainly to be seen. The pillar is of friable stone, in horizontal layers, and is constantly decaying away by the action of the weather. The top is formed by a stratum of hard crystalline rock, which resists the rain and sun, and is apparently now of the same diameter that the pillar which supports it originally was. (Wallace 1979: 99)

It is interesting to compare Wallace's description of *Pedra do Pilão* to its appearance today, over 120 years later (Fig. 4.1). Wallace did not appear to report any of the rock art that exists at *Pedra do Pilão* today, nor did he mention any of the graffiti. Perhaps he was not close enough to make that observation.

While hiking the length of *Serra da Paituna* from the base of the hill, Wallace's group ran out of water due to the increasingly oppressive heat and sun. His Indian guides led him to a hill spring to fill up their calabashes. The importance this signifies is that it describes how previous inhabitants living in the hills might have sustained themselves in the past, and it explains how they knew where to find such valuable resources. Regarding that matter, Wallace wrote:

> We soon came in sight of the spot, and a group of Mauritia palms [alternatively spelled Buriti, Boriche or Moriche], which always grow in damp places, as well as some patches of brilliant green herbage, gave us hope. On reaching the palms we found a moist, boggy soil, but such a slow filtering of water among the weeds that it took nearly half an hour to fill our gourd. Seeing a mass of green at the very base of the perpendicular rocks higher up, where the spring appeared to issue, we proceeded there, and found to our great joy, a little trickle

C. S. Davis, *Before Sunset*, Conflict, Environment, and Social Complexity,
https://doi.org/10.1007/978-3-031-93373-8_4

Fig. 4.1 Photos of *Pedra do Pilão* (top) taken up-close in 2011 and (bottom) from down the road in 2009

> of pure and delightfully cool water, and a shady place where we could rest and eat our lunch in comfort. (Wallace 1979: 100)

After refreshing themselves, Wallace and his crew continued until they reached the spot where their informant said he found the cave with inscriptions, but because he had only been there once before, he could not find the entrance. And after trying several leads, Wallace decided to retire for the day, to return the next day with a better guide. However,

> On our way back we passed by a high cliff, on which were some of the picture- writings I had so much wished to see. They were executed in a red tint, produced apparently by rubbing them in with pieces of the rock, which in places is of that color. They looked quite fresh, and were not at all obliterated by the weather, though no one knows their antiquity. They consisted of various figures, rudely executed, some representing animals, as the alligator and birds, others like some household utensils, and others again circles and mathematical figures, while there were some very complicated and fantastic forms, and all were scattered irregularly over the rock to the height of eight or ten feet. The size of most of the figures was from one to two feet. I took a general sketch of the whole, and some accurate tracings of the more curious single figures, which have unfortunately been since lost. (Wallace 1979: 100–101)

The cave that Wallace's guide was trying to escort him to was probably *Caverna da Pedra Pintada*, whose entrance is even concealed today by forest vegetation, save for the fact that now a dirt path brings a traveler right in front of it (Fig. 4.2).

During Wallace's time, no visible trail seems to have existed, at least not according to his description of the land from his vantage point atop *Pedra do Pilão* (Wallace 1889: 103). However, it is important to attempt to determine his exact location and route, as his descriptions of the rock art panel might refer to *Painel do Pilão*. His known whereabouts suggest this could be the rock panel he described, though he

Fig. 4.2 Entrance to *Caverna da Pedra Pintada* showing author taking panorama. Gigapan panorama of the cave can be viewed at http://gigapan.com/gigapans/195839

claimed it was perched on a high cliff, and that several of the drawings were one to two feet in length. *Painel do Pilão* features circles with connected vertical lines, which—according to some—might be interpreted as spoons. The outcrop itself overlooks a gradually descending slope of the hill all the way to the river, so the site could indeed appear to be perched high. However, from the present-day footpath, *Painel do Pilão* does not require much of a climb to reach (Fig. 4.3).

The next day, Wallace and his crew set out for the painted cave again—this time with their host, who took them by horseback but still could not find the entrance—and so they sent Indians to fetch an old man who knew the hills better. While waiting for his retrieval, Wallace's host—Sr. Nunez took him to Serra da Lua, where Wallace wrote his impression of the art there and compared it to the art he had encountered the previous day. Wallace writes:

> They [the rock paintings] were much larger than the others, and extended higher up the rock; the figures, too, were all different, consisting principally of large concentric circles, called by the natives the sun and moon, and several others more complicated and three or four feet high. Among them were two dates of years about 1770, in very neat well-formed figures which I have no doubt were the work of some travelers who wished to show that they knew how the others were executed, and to record the date of their visit. Near some of the higher figures were two or three impressions of hands in the same color, showing the palm and all the fingers very distinctly, as if the person executing the upper figures had stood on another's shoulders and supported himself with one hand (smeared with the red color) while he drew with the other. I also took copies of the figures at this place, which,

Fig. 4.3 *Painel do Pilão* site angled to capture the grid image in the foreground and the southeastern view overlooking the Amazon River in the distance

> being large and exposed, are visible from a considerable distance round, and are more generally known than the others, which are in a secluded and out-of-the-way situation, and were probably not visited by any European traveler before myself. (Wallace 1979: 101)

Wallace and Sr. Nunez returned to the group to find the local elder guide waiting for them; and so, they went to the cave entrance (Gruta Itatupaoca), of which Wallace wrote:

> The entrance is a crude archway, fifteen or twenty feet high; but what is most curious is a thin piece of rock which runs completely across the opening, about five feet from the ground, like an irregular flat board [a transom—a horizontal beam reinforcing the stern of a vessel]. This stone has not fallen into its present position but is a portion of the solid rock harder than the rest, so that it has resisted the force, which cleared away the material above and below it. Inside there is a large irregularly arched chamber, with a smooth sandy floor, and at the end there are openings into other chambers; but as we had not brought candles, we could not explore them. There was nothing about the cave at all remarkable, except the flat transverse rock at its mouth. The vegetation around it was by no means luxuriant or beautiful, nor were there any flowers worth noticing. (Wallace 1979: 102)

Although Alfred Wallace was a visitor to the area who spent no more than two days exploring, we get the impression that *Pedra do Pilão* and *Gruta Itatupaoca* did not have any notable rock art at the time. However, this should only be taken with a grain of salt because vegetative growth and other factors may have obscured Wallace's detection of some paintings. He does describe the top formation at *Pedra do Pilão* as being circular all around—like a mushroom (perhaps a flat-top mushroom)—instead of its present-day triangular shape, and we also get a consistent description of *Serra da Lua* compared to the present day, and a possible description of *Painel do Pilão*. Unfortunately, Wallace's recollection of a rock painting dated 1770 is not the same date of 1764 that exists in *Serra da Lua* today (Fig. 4.4). Perhaps Wallace was at an unknown location other than *Serra da Lua*, or he had not carefully recorded his observation, or he had simply rounded up the date, or he had entirely failed to see the earlier date. What we should, therefore, conclude from Wallace's account is that there were pre-existing rock paintings that, despite their "fresh" appearance, were indeed so old that none of the local inhabitants had any idea of their creation. These paintings were figures of the sun and moon, birds, animals, alligators, handprints, circles, mathematical-looking notations, and images that—to Wallace—resembled domestic utensils.

Wallace also wrote about calabashes (gourds) that were made in the Monte Alegre region (Fig. 4.5). These gourd bowls are important because they were often decorated with paints that were probably manufactured in similar fashion to the paints used in the rock paintings. Wallace wrote,

> The calabashes, or "cuyas," [cuias] are made in great quantities, and exported to Pará and all parts of the Amazon. They are very neatly finished, scraped thin, and either stained of a shining black or painted in brilliant colors and gilt. The designs are fanciful, with sometimes figures of birds and animals, and are filled up with much taste and popularity. The Indian women make the colors themselves from various vegetable juices or from the yellow earth, and they are so permanent that the vessels may be constantly wetted for a long time without injury. There is no other place on the whole Amazon where painted calabashes are made with such taste and brilliancy of color (Wallace 1979: 99)

Fig. 4.4 Colonial-period graffiti at *Serra da Lua*: (top) enlarged and Photoshop lab color A red enhanced photo. (middle) raw color photo. (bottom) Dstretch_rgb photo enhancement shows that the 1764 graffiti has a different spectral signature than the surrounding images, its pigment was either not made by the same technique or its paint was applied at a different time than the surrounding paintings. Note also that the handprint in the upper right corner is also darker but not as dark as the graffiti

A team led by Louis Agassiz traveled through the area in 1865. Although Agassiz wrote about the process of making paint and resin for *cuias* (traditional gourd bowls) that he observed an Indian woman from the village of *Surubiju* near *Serra da Lua* performing, he did not comment on the rock art of the hills (Agassiz and Agassiz 1868: 167, 361–164).

In fact, neither Agassiz nor Wallace was the first to write about the paint used on the *cuias*. Friar João Daniel, a Jesuit, wrote about the village of *Monte Alegre* after

Fig. 4.5 A *cuia* (gourd) that I cut, hollowed out, and attempted to paint with red (ochre) mud and red *Jutaí* tree roots while living in the hills of Monte Alegre

spending time there sometime between A.D. 1745 and 1757. Of the bowls, he wrote [translated from old Portuguese]: "The *cuias*, cups widely used by the Americans, and in truth, good cups by all estimates, rank as top priority; for the Indians know how to apply with such mastery the varnish and paint, that they never fade. [The process] had already stirred such curiosity that they [the Americans?] wanted to experiment with the bonding of this varnish, thinking it no different from some of the best lacquerware of China" (Daniel 1841: 434).

The friar's comments suggest a manufacturing process unique to the Indigenous people of the *Monte Alegre* region, which was likely applied to some (but not all) of the rock art in the distant past. This helps explain why some images still appear fresh, while others have faded due to exposure to rain or fire. However, we should not rule out the possibility that more recent generations applied or reapplied varnish to ancient images in an effort to preserve them. As Wallace noted, someone in 1764 appears to have duplicated the painting process, as evidenced by the painted date, which still exists today. Images that did not receive such preservative treatments, however, seem to have not lasted as long.

That said, I want to caution against assuming that the 1764 graffiti was made in the same manner as the other rock art. The *Dstretch* color enhancement reveals a different tint of red in the graffiti compared to the surrounding rock art, suggesting a distinct method or material was used. By the mid to late eighteenth century, colonial European countries had oil-based and lead-based paints (Whitehill 2012), and this open-air date may have been applied using one of those paints.

In contrast, I discovered another early graffiti date of 1721 in the *Gruta 15 de Março* cave. It was accompanied with what appears to be the word "Juex," a probably French or Dutch word (*jeux* or *jeu*) with uncertain meaning—perhaps *toy* or *game*. The graffiti was red and yellow in color, like the native paints, but it was so

Fig. 4.6 Photo of wall in *Gruta 15 de Março* taken in 2010 and enhanced with Dstretch_lab_ac to enhance all colors. It appears to display "Juex 1721." This date predates all known settlements in this region. A Portuguese map from A.D. 1623 shows that the Flemmings had been to the region at the juncture of the Gurupatuba and Maicuru Rivers to trade with the Tapajos Indians, and a map from A.D. 1630 shows a Dutch fort was erected at the mouth of the Maicuru, but it had been destroyed by the Portuguese—presumably around A.D. 1623

severely eroded that I only discovered it while analyzing photos with Dstretch image enhancements. This date, combined with the foreign word, more strongly suggests it was made by colonizers. It was in a cave, protected from harsher weathering, and yet it was far more eroded (Fig. 4.6). Certainly, the native painting method was not known at this time, and better-quality European paints were not yet applied here either.

Charles Frederick Hartt

In 1870–1871, the geologist Charles Frederick Hartt, who was part of Agassiz's previous group, returned with the Morgan expedition and recorded the rock art at *Serra do Ererê* in more detail. His description of many of the images led him to the conclusion that "the most important [images] among these appear to be representations of the sun, moon, and stars" (Hartt 1871: 142). Hartt further explains, "I think that these figures point to a worship of the sun by the tribes which executed them … and in the vicinity of the rock tower at Ererê, seems to me that those places had something of a sacred character and were often resorted to" (Hartt 1871: 147). The "Comet-Eclipse panel" (see Fig. 4.7) is the closest panel of images to the rock tower (Fig. 4.8).

Fig. 4.7 Labeled as the "Comet-Eclipse" panel for what they might possibly represent, these paintings are discussed in greater detail in Chap. 6

Charles Hartt's description is most important, because although Alfred Wallace visited the area first, Charles Hartt seems to have studied and copied the paintings in great detail. His account and impression of the rock art near the rock tower of *Serra da Lua* are so important that a full paraphrase of his descriptions was published in the 1891 London publication (which had no listed author), *Archaic Rock Inscriptions* (Jennings 2003) included here:

> On these walls of rock, at and near the western end of the Serra, sometimes near their base, sometimes high up in conspicuous situations difficult of access, are great numbers of rude characters and figures, for the most part in red paint, some isolated, others in groups. Some rock surfaces are thickly covered with them, many being so washed by rains and defaced by fires as not to be traced out, others being brought out fresh, suggesting that they were not all executed at the same time. Standing just in advance of the line of cliffs at some distance east of the western end of the Serra, is a tall, tower like mass of sandstone painted not only on the base, but high up on the sides, while the cliffs behind and on both sides are covered with figures. All the localities are very conspicuous, and some of them are so large as to be visible at the distance of more than a mile. (Jennings 2003: 57–58) Professor Hartt says he is disposed to think that they [the large concentric circles] are intended to represent the moon, since they are not furnished with rays. One figure on the cliff at the western end of the Serra,

Fig. 4.8 The rock tower that Charles Hartt references as having a sacred sun-worship character to it. This photo was taken with a zoom lens from the base of the hill. The rock tower is discussed in much greater detail in Chap. 6

> undoubtedly represents this heavenly body. Besides the above forms there are rayed figures in abundance. Sometimes they consist of a single circle or several concentric circles, the outer, one, only, being rayed, but on the side of the great rock on the top of the Serra is a figure a foot in diameter, and very distinct, formed of two concentric circles, each with a few large tooth shaped rays. Part of this figure is obliterated. At the same locality is another figure consisting of a circle with serrated rays with only a spot in the center. There are also circles, single and double, sometimes nucleated, which bear rays only on the upper side; and rayed spirals, some of which appear to represent stars, some of which are drawn, and others engraved. Some of the figures are rude and grotesque in the extreme; one, about three feet six inches high looks like the sun with a human body; another at the western end of the Serra is a curious rayed head, ornamented at the top with what looks very much like a comet. Animals of various kinds are also depicted as well as the human form and the sun, moon, and stars. At the west end of the Serra is the figure of a snake; in another place is seen what the Indians call a *mucura*, a species of opossum, and others supposed to represent alligators. Birds are very seldom seen, a few markings only which may be taken for such being found. Several drawings of the sea cow exist, and a few of fishes; none of the dog, ox, or horse. A very large number of the cutting and paintings are too mystical to give any meaning whatever to, and appear like the productions of mere children. (Jennings 2003)

From the 1891 publication we get a description of some paintings that are easily identifiable with those that exist today (Fig. 4.9). The publication quotes Charles Hartt's conclusions:

> Professor Hartt says :—"The antiquity of the rock paintings and sculptures of Eastern South America is undoubted, and they are mentioned by many of the ancient writers, as well as by Humboldt, and others in more recent times. It is well known that the drawings of Ererê and

Fig. 4.9 Probably Hartt's description of (left) two concentric circles with large tooth-shaped rays at *Mirante* in 2007, (center) false color enhancement of the curious-rayed head ornamented at the top with what he thought was a comet, and (right) a Dstretch lrd enhanced photo of a probable *mucura* opossum that Hartt describes beyond the rock tower at *Serra da Lua* in 2010

> those of Obidos existed more than two hundred years ago. There can be no doubt that they antedate the civilization of the Amazonas, and there is a strong probability that some of them, at least, were drawn before the discovery of America. I hold it most probable that the rock paintings and sculptures were made by tribes that inhabited the Amazonas prior to the Tupi invasion. The sculptures are supposed to be older than the paintings." (Jennings 2003: 59)

Hartt's conclusions about the antiquity of the rock art were further supported in the publication by similar quotes from a Senhor Pena, who stated:

> I know of no trace of sun-worship among the uncivilized Indians of Pará to-day, nor do they make rock paintings or inscriptions. The greater part of the Brazilian Indians, such as the Tupis, Botocudas, etc., appear to have had no idea of a God, and no form of worship. We have no historical account of the practice of sun-worship among the ancient Indians of the Amazonas. The probabilities are, that the tribes anciently inhabiting the Amazonas were more advanced in religious ideas than those Brazilian Indians of which history gives us an account. (Jennings 2003)

While some of Pena's statements are not entirely accurate (most Indigenous groups had forms of ancestor or animism worship, and some groups—particularly those living north of the Amazon River, including certain Tupi groups and even the legendary Amazons—did have a reverence for the sun and/or moon, as discussed in Chap. 10), he and Hartt reached the same conclusion as this research. Most of the art at *Monte Alegre*, particularly in the vicinity of *Serra da Lua,* appears to serve as a solar observatory.

Shortly after his Amazonian expedition, Charles Hartt became an integral figure in the Brazilian Geological Commission. He initiated several other expeditions through the region to study the geological formations of the *Monte Alegre Dome* (Hartt 1896). Herbert Smith conducted one such expedition to explore the Gurupatuba drainage system, which feeds into the Amazon River at *Monte Alegre*. Through his travels, Smith described the Indians in the village of Ererê, noted the endurance of some of their traditions (Smith 1879: 346, 371–397), and summarized regional myths shared by previous explorers, including the Tupi belief that all things were created by either the sun, the moon, or a third water deity (Smith 1879: 586, 541–587). In 1924, Curt Nimuendajú copied paintings at *Serra da Lua, Pedra do Mirante,* and *Serra do Sol in Ererê*, as well as paintings at *Gruta do Pilão* (also known as *Pedra Pintada*) in *Serra da Paituna* (Pereira 2002).

Katzer and Germany

In 1933 Frederich Katzer, a geologist, surveyed the region and wrote about the existence of rock art in the area. According to Katzer, in the sector north of *Serra do Ererê*, there is [translated from Portuguese] "a large, isolated, sandstone block, with a smooth surface, that had been covered with indigenous inscriptions and designs" (Katzer 1933: 115). This location is identified today as *Pedra do Mirante.* In 1954 and 1955 another German scholar, Manfrend Rauschert, surveyed the region and wrote about the rock art of *Serra da Lua* and *Serra do Sol* (Milhomens et al. 2008: 155).

Recent Research

More recent archaeological expeditions were conducted in 1984 by the Grupo *Espeleológico do Pará* (Silveira et al. 1984) and in 1986 by the Uruguayan archaeologist Mario Consens, who assessed the archaeological potential of the region (Consens 1989). In 1996 and 2002, Edith Pereira made further significant discoveries and descriptions of images across 14 different archaeological sites, providing the most comprehensive catalog of the rock art in the *Monte Alegre* region at the time (Pereira 2002).

In 1991–92, the first stratigraphic research was conducted at *Caverna da Pedra Pintada,* located in *Serra da Paituna,* by Anna Roosevelt and her team (1996). The Roosevelt excavations yielded paint drops and lumps of pigment found in stratigraphic layers dated between approximately 13,700 and 11,800 years BP, based on 53 radiocarbon dates from plant remains and sediment associated with Paleoindian artifacts, black soil, and other food remains. The deposited paint samples were chemically consistent with the pigment used on the painted walls above the excavation.

Further supporting evidence included some paintings located low on the wall surface today, which would have been at a more optimal viewing level in relation to the Late Pleistocene cave floor before the subsequent layering of Holocene sediment (Roosevelt 1999: 376–379; Roosevelt et al. 2002: 26; Roosevelt et al. 1996: 190–192). Since only one pigment fragment was found in the later Holocene layers, Roosevelt concluded that much of the rock art was created during the Late Pleistocene. Additionally, ceramic-period layers of dark gray sand, radiocarbon-dated between 7500 and 1450 BP (Roosevelt 1999: 376–379; Roosevelt et al. 2002:

26; Roosevelt et al. 1996: 190–192), were separated from the more ancient Paleoindian layers by a sterile, tan-colored sand layer.

References

Agassiz, L., and E. C. C. Agassiz. 1868. *A Journey in Brazil*. Boston, MA: Ticknor and Fields.

Consens, M. 1989. Arte Rupestre No Para: Analise De Alguns Sitios De Monte Alegre. *Dedalo* 1 (especial): 265–278.

Daniel, F. J. J. 1841. Parte Segunda Do Thesouro Descoberto No Rio Amazonas- Noticia Geral Dos Indios Seus Naturaes, E De Algumas Nações Em Particular: Da Sua Fé, Vida, Costumes, E Das Cousas Mais Notaveis Da Sua Rusticidade. *Jornal do Instituto Historico Geographico Brasileiro* 3 (12): 422–441.

Hartt, C. F. 1871. Brazilian Rock Inscriptions. *American Naturalist* 5 (3): 139–147.

Hartt, C. F. 1896. A Geologia Do Para. *Boletim do Museu Paraense Emilio Goeldi de Historia Natural e Etnografia* Tomo 1 (Fasciculos 1–4): 257–273.

Jennings, H. 2003. *Archaic Rock Inscriptions; an Account of the Cup and Ring Markings on the Sculpted Stones of the Old and New Worlds*. London: Kessinger Publishing.

Katzer, F. 1933. Geologia Do Estado Do Pará. In *Boletim do Museu Paraense Emílio Goeldi de História Natural e Etnografia*, ed. M. P. E. Goeldi. Belém, Brazil.

Milhomens, A., D. S. Nascimento, L. Salvo, W. Rodrigues, F. Ferreira, A. Mota, R. Gangana, E. R. de Carvalho, and R. d. C. M. Moreira 2008. Plano De Manejo Do Parque Estadual Monte Alegre (Pema), ed. P.-P. d. D. d. E. n. A. Legal, 303. Ministerio do Meio Ambiente, Belem.

Pereira, E. 2002. *Arte Rupestre Na Amazonia Para*. Belem: Museu Paraense Emilio Goeldi.

Roosevelt, A. C. 1999. The Development of Prehistoric Complex Societies: Amazonia, a Tropical Forest. *Archeological Papers of the American Anthropological Association* 9 (1): 13–33.

Roosevelt, A. C., J. Douglas, and L. Brown 2002. The Migrations and Adaptations of the First Americans: Clovis and Preclovis Viewed from South America. In *The First Americans: The Pleistocene Colonization of the New World*, ed. N. G. Jablonski, 159–236. vol. 27. continuing vols. Berkeley: Memoirs of the California Academy of Sciences.

Roosevelt, A. C., M. Lima da Costa, C. L. Machado, M. Michab, N. Mercier, H. Valladas, J. Feathers, W. Barnett, M. Imazio, A. Henderson, J. Silva, B. Chernoff, D. S. Reese, J. A. Holman, N. Toth, and K. Schick. 1996. Paleoindian Cave Dwellers in the Amazon: The Peopling of the Americas. *Science* 272 (5260): 373–384.

Silveira, L. T., R. V. L. Pinheiro, and S. V. L. Pinheiro. 1984. *Roteiro Espeleologico Das Serras Do Ererê E Paituna*. Belém, Brazil: Inedito.

Smith, H. H. 1879. *Brazil the Amazons and the Coast*. Sampson Low, Marston, Searle, and Rivington, London.

Wallace, A. R. 1889. *A Narrative of Travels: Amazon and Rio Negro, with an Account of the Native Tribes*. 2nd ed. London: Ward, Lock and Co.

Wallace, A. R. 1979. *Viagens Pelos Rios Amazonas E Negro*. 2nd ed. Translated by E. Amado. Coleção Reconquista Do Brasil 50. Livraria Itatiaia Editora LTDA, Belo Horizontes.

Whitehill, W. M. 2012. *The Arts in Early American History: Needs and Opportunities for Study*. UNC Press Books.

Chapter 5
Serra Da Lua and Serra Do Sol Rock Art

Serra da Lua is the name of the westernmost location on *Ererê* hill (Fig. 5.1). This location has rock art (pictographs) spread out alongside a rock and dirt trail halfway up the hill. The trail begins after a scamper (now there are wooden steps installed) of about 72 m up the hill, which has a total elevation of ~117 m above sea level (masl). The trail path mostly hangs close to the wall, with relatively gentle rises and dips. The rock art is clustered in some places (called panels) but dispersed in between the 8-or-so panels. The worn path terminates at a rock tower (Fig. 5.2)—the same rock tower mentioned by Charles Hartt and Alfred Russell Wallace.

Beyond the rock tower, a much fainter path continues through thicker rainforest growth further along the hill's perimeter. A few pictographs are sparsely strewn along this path, but it is easier to miss them due to the overgrowth of vegetation. The path (and vegetation) clears again, and two cliff wall panels become visible at a site called *Serra do Sol* (Fig. 5.3). This site is along the northwestern side of *Ererê*.

Serra do Sol is a more discrete location of two main wall panels separated by about 74 m, with a few pictographs painted between the span. *Serra da Lua* spans a much longer path, approximately 266 m from the beginning of the sequence to the rock tower (Fig. 5.4). There are also a few paintings beyond the rock tower *en route* to *Serra do Sol*. However, at *Serra do Sol*, there is also a small, fist-sized hole in the rock where a cache of lithic artifacts was found. Many of these cached lithics appear to have been partially heated or burned (scorched) on half or part of their sides. They are described later in this chapter.

Rock Art Motifs at Serra da Lua

There are numerous images at *Monte Alegre* that span a range of themes. Earlier scholars categorized them into five common motifs: "anthropomorphs, hands, zoomorphs, biomorphs, and pure graphs" (Pereira 2002; Silveira et al. 1984). The fact

C. S. Davis, *Before Sunset*, Conflict, Environment, and Social Complexity,
https://doi.org/10.1007/978-3-031-93373-8_5

Fig. 5.1 Serra da Lua western cliff at the southern start of the pictographic rock art trail, which curves around Ererê hill to the north. Photo taken 6:24 PM on August 19, 2023

Fig. 5.2 The Rock Tower at the terminus of the *Serra da Lua* pictographic trail is the free-standing structure visible to the far-left side of the photo (taken at 6:02 AM Nov 16, 2007)

that the art can be categorized with any degree of regularity supports the idea that it carries site-specific significance and uniqueness, beyond being mere doodles or happenstance drawings. Some of the art mimics features characteristic of the region, such as the red and yellow lichens and the sun and moon halos. For these images, the art may not only represent perceived reality but also serve as a type of landmark

Fig. 5.3 *Serra do Sol* as seen from the far-distant road curving around the north side of Ererê hill. A faint red dot can be seen left of center in the photo, which is the northern solstice concentric pictograph discussed in the *Serra do Sol* section

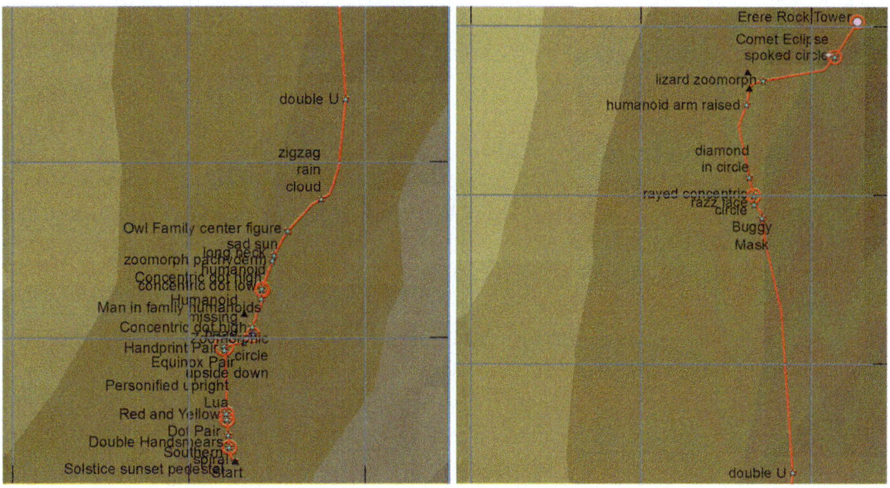

Fig. 5.4 Serra da Lua Pictograph Trail. This foot trail along the cliff of the hill begins at the bottom of the image (south) and terminates at the rock tower mentioned by Charles Hartt at the top of the image (north). The trail preserves distance and general cardinal direction at scale, but it does not show the exact cardinal directions of each pictograph

identity. Alternatively, we should consider whether the rock art might have functioned as personal identifiers or commemorations, with the artist marking the statement, "I was here," or "I come from the land of red and yellow lichen."

I added a few more motif categories and revised some of the previous five motifs of past scholars for more comprehensive and analytical comparisons. Categorization has revealed a trend among images at *Serra da Lua* (Table 5.1). Other sites also appear to fit trends—for example, most of the caves have a higher percentage of zoomorphic paintings, and all open-air sites have a higher percentage of rayed or concentric circles and dots. However, only *Serra da Lua*'s rock art stretches over a

Table 5.1 Serra Da Lua Motif Analysis Summary Statistics

Motifs	Count
Abstract	34
Altered orifice	3
Arc	3
Circle	29
Dot	58
Eye	4
Face	33
Fireball/comet	4
Geometric	30
Hand smears	2
Handprint	33
Humanoid	11
Meander	3
Logogram	6
Spiral	4
Zoomorph	26
Total	283

long enough distance to allow subdivisions of panels by themes. More on this will be discussed in the section titled *Panels*. It also provides the longest collection of rock art images by which to establish categories that other sites in Monte Alegre fit only a subset. The comprehensive set of motifs are as follows.

The "Dot" motif (Fig. 5.5) is assigned to filled-in circles. The diameter of the dots ranges, but generally, they fit any diameter that requires the act of painting inside of a circle (as opposed to just dabbing or punctuating a single point). They are the most numerous designs, particularly at Serra da Lua. Dots are not always perfectly circular, but they are represented alone or in combination with other designs, particularly circumscribed by circles, bands, or rays projecting from them. Some rays radiate straight out from a dot or a circle around a dot; other rays curl as they radiate out. Instead of rays, some dots can have a single line radiating out from them or two stems resembling horns or antennae.

Circles are perhaps just as numerous as dots when considering the number of circles that circumscribe other forms, or that compose a face or torso. The category of circle here, however, refers to a design with a simple circle at its center (Fig. 5.6). Nearly half as many circles than dots are circumscribed by other circles. However, some of these circles are circumscribed by more than one other circle. Together, circles and dots, and those that are circumscribed or rayed, are the most numerous designs at *Serra da Lua*. They are also found at other Monte Alegre rock art sites.

There are two styles of spirals at *Serra da Lua*. One spiral type includes a line that coils in one direction around itself so that the overall form resembles a whirlpool spinning in or out. The other type of spiral (Fig. 5.7) resembles an S and resembles a cornu spiral. One end of the spiral coils around itself once or twice and then extends out for a distance before coiling in the opposite direction. S-type cornu

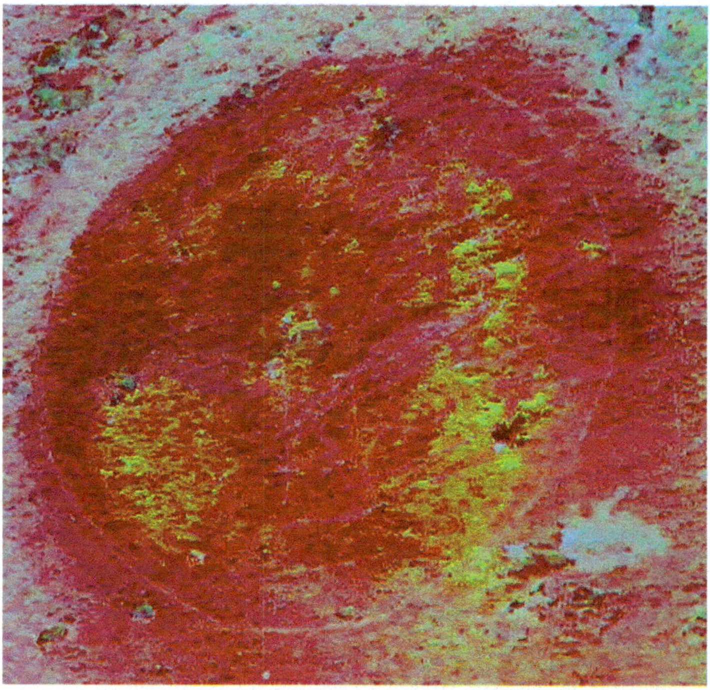

Fig. 5.5 Dot motif, RGB color enhanced to make red and yellow more vibrant. A faint circle was scratched on the wall but red paint overlays some parts, suggesting the scratched circle was made first, then yellow paint filled in the circle, followed by red paint filling over much of the yellow. The red largely borders the scratched circle, perhaps as an attempt to make concentric bands of red and yellow

spirals often lay on their side or at an angle. Both types of spirals are rare, but when they appear, they seem to be placed strategically, often low on the wall and either at the beginning of a sequence of paintings or at the end.

Humanoids (Fig. 5.8) are drawings that resemble full human figures, typically with a head, torso, arms, and legs. The torso is usually depicted as a simple stick figure, with the body in a forward-facing position and the limbs splayed outward, bent at the elbows and knees at right angles. The face is often represented by points or circles for eyes, a horizontal arc or line for the mouth, and usually lacks any indication of a nose. Faces typically convey a facial expression, such as a smile or frown.

Additional lines are sometimes shown on the forehead. Elements indicating female sex, such as hair or breasts, are generally absent, with some debatable exceptions. Image enhancements from certain photos, however, have revealed humanoids that appear to depict male genitalia. While these features were previously interpreted as tails, menstrual bleeding, or babies, the enhanced images from Dstretch clearly show ovals on either side of a vertical line, suggesting male genitalia.

The humanoid category largely corresponds to Silveira's "anthropomorphs" (Pereira 2002; Silveira et al. 1984).

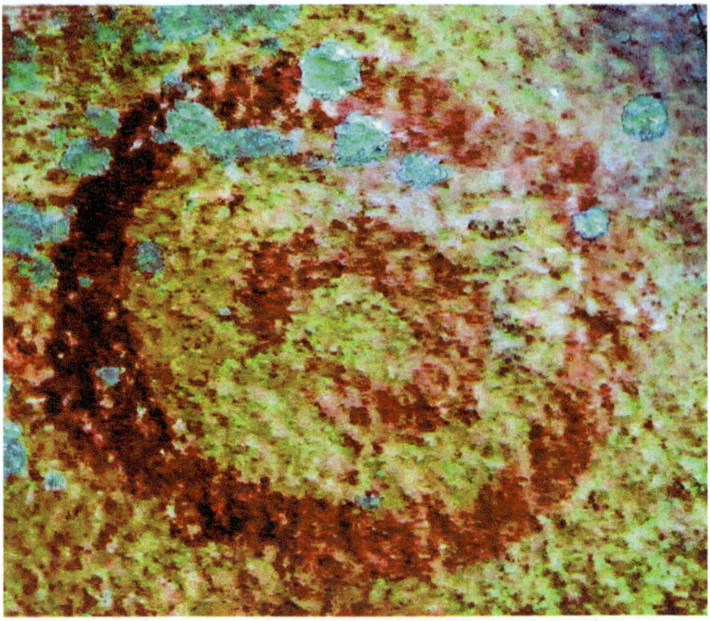

Fig. 5.6 Circle motif. Depicted is an inner circle surrounded by an outer circle—a circumscribed circle. This photo is color enhanced to make red and yellow more vibrant

I separate disembodied faces into their own motif, separate from the humanoid motif. Faces would have been included in Silveira's "anthromorphs." Some faces are depicted with features bounded by a usually circular head, while others are unbound groupings of facial features: eyes, nose, mouth (Fig. 5.9). A disembodied face tends to show more expression than faces on full body humanoids, but faces sometimes have linear features extending from them. Two such face images, one at *Serra da Lua* and the other at *Gruta dos Ossos II*, seem to personify comets. Other faces seem to depict animals or ambiguous features. These images might reference an animal spirit, a ceremonial mask, painted faces, group or individual identity characteristics, or other ritual associations.

Eyes also would have been included in Silveira's "anthromorphs," but I distinguish floating eyes from humanoids and faces. They are much less common, but a few eyes appear in isolation from an entire face (Fig. 5.10). Where distinct eyes appear, they tend to have more detail (pupil, iris, eyelid, eye lashes), and they seem to be in close proximity to other images, perhaps to signify that the painting nearby was something seen personally. A few other eye images are more circular and resemble the sun or moon, or other circular features blended with an eye.

Handprints are paintings created by wetting a human hand in paint and then pressing it against the wall, but there are also a few cases where the entire image of a hand was drawn to resemble an imprint. Many handprints also have circles or geometric designs drawn in the palm (Fig. 5.11). Most handprints are slightly to much smaller than the size of an adult male hand today, suggesting that the artists

Fig. 5.7 Spiral motif. Depicted above is a spiral resembling a cornu spiral. The more traditional "whirlpool" spiral is also found among the rock art at Monte Alegre, but less so. This photo is color enhanced to make red more vibrant

were either small-statured adult males, females, or adolescents. Additionally, there are a few yellow handprints at Painel do Pilão that appear to correspond to the hands of very young toddlers. These prints are somewhat unconventional, as the ratio of palm width to hand length seems larger than most human hands today. However, it's possible that the width of the handprints was increased by rocking the hands from side to side against the wall.

Both right and left hands are represented in the rock art, though typically not in pairs. Instead, groups of left hands are found in one location, and groups of right hands in another. When handprints do appear in pairs, they seem to have been made by the same person. In some cases, a right- and left-hand pair may have been intentionally overlapped.

Additionally, there are handprints that consist only of the fingertips, usually from the most distal two segments of the largest three fingers. It's unclear whether these marks were intentionally made or if they resulted from the individual bracing

Fig. 5.8 Humanoid motif. Depicted above are three humanoid stick figures. Two are visible with heads, the third is between the two, and no head is visible. Also, its limbs extend at a diagonal instead of bent at right angles. The hands of the other two figures are unconventional

themselves against the wall for balance. Some handprints show bent or curved fingers, suggesting the person may have suffered a fracture or dislocation of one or more fingers. There are also two instances of partial footprints in red paint, although these may have been made unintentionally.

Zoomorphs are designs that resemble animals, with the most common being lizards, snakes, monkeys, manatees (Fig. 5.12), fish, and larger birds such as owls and vultures. A few designs are more difficult to identify, such as a possible possum, and perhaps a jaguar. These animal figures are generally drawn in a similar stick-figure style to the humanoids, except that their torsos are typically polygonal in shape rather than linear. This zoomorph motif encompasses two of Silveira's categories—zoomorphs and biomorphs (Pereira 2002; Silveira et al. 1984).

Sometimes, zoomorphs are depicted from a side view, with their limbs positioned accordingly. However, there are exceptions to this rule, including figures that may either represent humans or tree sloths. These animals are common in the region and bear similar proportions to humans, except for their claws. This ambiguity is difficult to resolve, as most zoomorphic images are isolated with no contextual

Fig. 5.9 Face motif. Depicted above is a color-enhanced photo of a face resembling a human, with half arc pupils and points as eyeballs. A midline between the eyes reflects the bridge of the nose and a downturned arc reflects a frown. A short red bar extends down from the frown, perhaps indicating a tongue. Above each eye is a pair of lines resembling eyelashes. This photo was color enhanced to make the red color more vibrant

clues, such as their habitat (e.g., in a tree or underwater). Consequently, the default approach is to categorize most ambiguous figures as zoomorphs, modified by the geometry of their torso (diamond, rectangular, oval, etc.). Some zoomorphs combine humanoid and animal features. These mixed figures are referred to as therianthropes.

The geometric motif (Fig. 5.13) is attributed to isolated geometric shapes besides circular features and correlates to Silveira's category of pure graphs. However, even among the geometric motifs, not all paintings are single geometric shapes; they can also consist of several geometric elements combined to make complex designs. The most common elements used to compose the designs are vertical, horizontal, or diagonal lines, zigzags, arcs, and various polygons. Also, other geometric designs are distinct enough to be designated astronomical designs, like planets or stars, due to context—especially when they are near similar sky-themed designs like concentric circles and dots. This line of reasoning is used later to claim the depiction of eclipses and comets.

Fig. 5.10 Eye motif. Depicted above is a "floating" eye (an eye in isolation of a face and/or body). This eye has an oval shape, an oval pupil, indications of the nictitating membrane in the left corner, and perhaps eye lashes extending from the right corner

Meanders are linear features that do not seem to have a consistent geometric pattern. Instead, they seem to wander across the wall surface (Fig. 5.14). In some cases, they could be snakes or worms, but there are several snake pictographs, and they typically have a defined "head." I suspect that meanders are directions for journeying, or they represent some feature of the landscape—either the horizon, or the outline of the rocky outcrop itself.

Meanders are rare, but when they appear, they are often near other motifs, such as circles or dots.

Logograms refer to designs that appear to consist of repetitive patterns, resembling a form of language or script, rather than naturalistic representations of observable objects (Fig. 5.15). This category was introduced after the use of the Dstretch algorithm, which enhanced photos of the paintings and revealed additional details in some pictographs. All images categorized as logograms share a common structural feature: they display a decreasing probability of repeated symbols, although each image remains visually distinct.

While this category is somewhat ambiguous, it is the most fitting description for the patterns observed within these images and their similarity to one another. Logograms have been identified above the cave opening of *Caverna Pequena* and at the three most painted sites in the region: *Serra da Lua, Painel do Pilão,* and *Caverna da Pedra Pintada*. However, due to the faintness of these images, and the

Fig. 5.11 Hand motif (handprints). Photographed above is a right-side human handprint with a whirlpool spiral, or a concentric point, drawn in the palm

fact that I discovered most of them only through photo enhancement, it is likely that additional logograms remain undetected at other sites as well.[1]

The "abstract" motif (Fig. 5.16) is a default category to group images that are indeterminable. Images that fall into this category might include accidental marks on the wall, paintings that have mostly eroded away, or drawings that have ancient cultural significance that is unknown or even unknowable to us today. Unlike the logogram category, images in this group do not have a sequence of symbols (see footnote on *Information Theory*), nor does one image here necessarily share any similarity to other images in this category. However, the abstract category is one of the more frequent categories, which some might interpret to indicate the antiquity of the images.

Altered orifices are the only motif that do not apply to painted pictographs. Nor are they petroglyphs. Orifices refer to fissures and cracks in the wall that show signs of intentional human modification to widen, smooth, and make the opening more symmetrical or aesthetic (Fig. 5.17). In some cases, ancient artists have further

[1]This definition comes from *Information Theory* which, for linguistic applications, identifies an intelligible language as such when a graph of its distinct sounds or symbols form a 45° slope (Shannon and Weaver 1949; Manning and Schütze 1999).

Fig. 5.12 Zoomorph motif. Depicted above is a pictograph considered to represent a manatee, due to the human-like forelimbs, but the fishtail hindlimb. The torso is a diamond shape in contrast to the stick torsos of humanoid motifs. The red dot positioned on the lower torso has unknown significance. The image is located high on the rock tower, however, so it could indicate a "sky" manatee—perhaps a constellation

highlighted these orifices by tracing their openings with paint. Other orifices have paintings inside them, while others have paintings positioned nearby with themes that refer to the openings. In a manner of speaking, orifices can be considered petroglyphs, but in truth, most altered orifices seem to have only had their openings enlarged or smoothed by an ablative process. Monte Alegre has very few true petroglyph engravings.

Fig. 5.13 Geometric motif. This zigzag is just one type of geometric shape included in this motif, but there are at least two specific zigzag pictographs at Monte Alegre. They have no clear implications to their meaning or significance, however

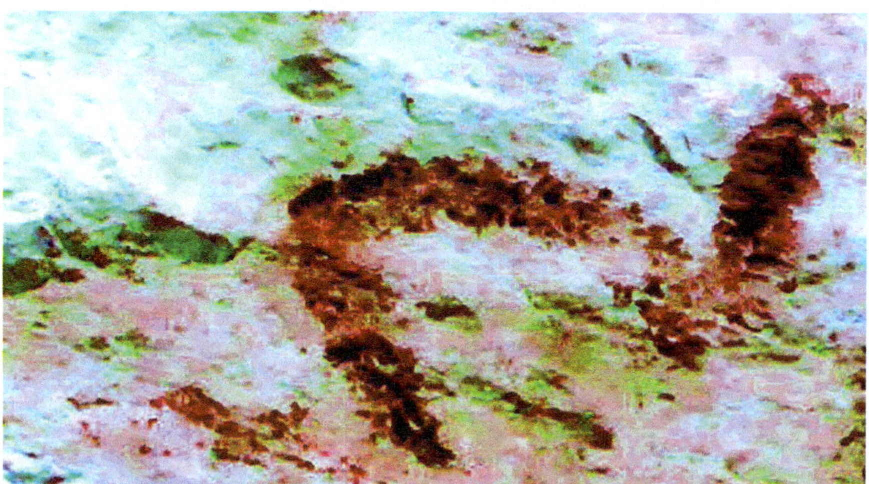

Fig. 5.14 Meander motif. Depicted above is a red curvilinear feature beneath a red point. The line has no specific pattern repeated elsewhere, and therefore it is designated as a meander. This photo was color enhanced to make the red color more vibrant

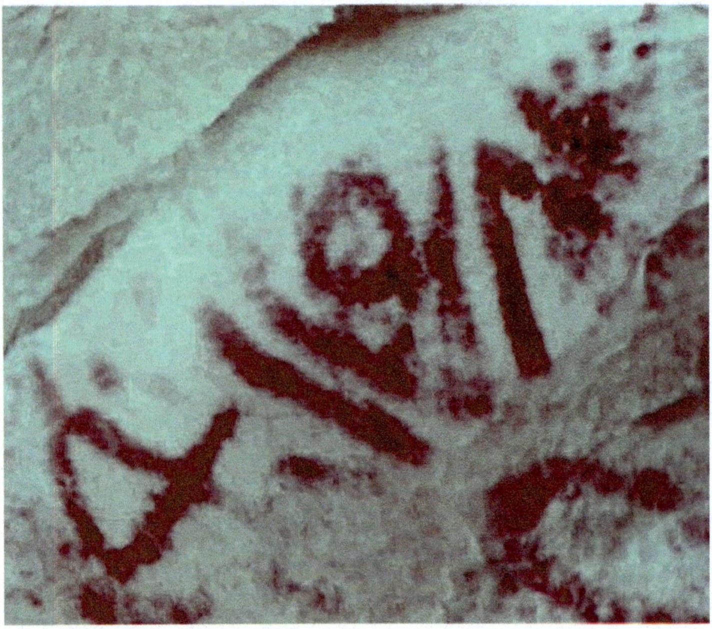

Fig. 5.15 Logogram motif. Depicted above is a sequence of marks on the rock tower at Serra da Lua. These symbols have no clear correlation, other than being painted in a linear sequence. These images could therefore represent some sort of ideogram or spoken language, however unlikely that seems for (possibly) 13,000-year-old Amazonian Paleoindians

Fig. 5.16 Abstract motif. Depicted above is a painting of a complex design at Serra da Lua that does not fit any of the prior motif categories. Any meaning, significance, interpretation, or association is currently unknown. This photo was color enhanced to make the red color more vibrant

Fig. 5.17 Orifice motif. Several holes in the walls at Serra da Lua (and a few other places) were widened, smoothed, and shape into geometric shapes for reasons unknown. Some holes may have held wooden posts, perhaps as scaffolding. However, some orifices have pictographs inside, like a smiley face or a concentric circle. This orifice has some faintly visible red marks near the center of the photo

Panels

For this book I use "panels" to refer to the location of a specific wall that has numerous paintings in proximity and in greater frequency than most other areas. The paintings are often arranged in a linear or rectangular sequence. No set number of paintings or a set distance of separation between them is specified, but panels are recognizable because they appear on the same wall, serve as landmarks along the foot trail, and in some cases, because they seem to contain a dominant theme. Some panels are a clutter of images, others are sparse (Fig. 5.18). All panels were photographed in high resolution using a Gigapan panorama machine that stitches together zoomed-in photographs to create a large (but artificially curved) canvas of the walls.

At *Serra da Lua* and Serra do Sol, I defined eleven different panels. For the rock tower, and a few other locations, I further subdivided panels into more densely painted regions and themes. For example, the rock tower has one major panel, but six different locations along its sides. Four locations contain concentric dots or circles, strongly suggestive of sky themes (sun most likely, but perhaps moon too). In fact, out of the 283 images tallied at *Serra da Lua* for this research, 95 of them (33%) are dots, circles, comets/fireballs, or spirals indicative of sky themes

Fig. 5.18 Gigapan panorama of the "Here I Am" panel at Serra do Sol. The panel is named for the upright figure on the lower left side of the photo with arms rays high. There is also a snake-like image, a disembodied face resembling an owl in the center, and at least two humanoid-like figures laying on their sides. Their heads point in the direction of the other panel at *Serra do Sol*, the Northern Solstice panel

Table 5.2 Serra Da Lua Motif Analysis Summary Statistics

Motifs	Count
Abstract	34
Altered orifice	3
Arc	3
Circle	29
Dot	58
Eye	4
Face	33
Fireball/comet	4
Geometric	30
Hand smears	2
Handprint	33
Humanoid	11
Meander	3
Logogram	6
Spiral	4
Zoomorph	26
Total	283

(Table 5.2). A sky theme was also designated for the two concentric circles in the Stick Family panel (Fig. 5.19). The rock tower will be discussed in detail in its own section.

Fig. 5.19 Gigapan panorama of the Stick Family panel, labeled due to the humanoid stick figures in the far-right corner of the stitched image. Due to this angle, the third (child-sized) stick figure is barely visible. Also, note the two concentric dots on the left side of the image. Their compass angles have been remeasured more recently (2016) with more precise instrumentation, which updates some previous reports

The Diving Figure panel has four different concentrations even though they are all on the same wall next to each other: the red and yellow pair, the two large concentric figures, the numerous handprints, and the fireball with graffiti. Focus is usually given to the concentric figures, so image enhancements help bring out missed details with the handprints. Categorizing the panel motifs (Table 5.3) has revealed that most handprints are found near the beginning of the trail, on the wall between the large "Red and Yellow Pair" and the "Diving Figure" (see next chapter). Most handprints here do not overlap the larger panel paintings, but they do sometimes overlap each other, and they are often positioned lower on the wall or in the wall spaces between the circles and dots. Many handprints (Fig. 5.20) are also faded much more than the major panel imagery, and some of the patterns drawn into their palms are the spiral motifs observed as other paintings at Monte Alegre, but this is uncertain because handprints are already very faint, and their details are often blurred or smeared. I did not find any handprints among the humanoid or zoomorphic panels, including the distinct and large "Owl Mates" panel, which features three owl-like figures. In fact, handprints do not reappear along the *Serra da Lua* foot trail until reaching the "comet-eclipse" panel (discussed in the next chapter) and the rock tower. Given the potential sky-related themes associated with the remaining panels, they will be addressed and displayed in the following chapter.

Table 5.3 Serra Da Lua–Serra Do Sol Panel Classification

Serra da lua panel analysis	Total paintings	Altered orifices	Handprints			Eye	Dot / circle					Fireball/ comet	Face/ mask	Humanoid	Therianthrope	Zoomorph	Geometric	Spiral	Meander	Logogram	Abstract
			Left	Right	Unknown		Total	Concentric	Crisscross	Rayed	Stemmed										
Sun pedestal	24	0	4	0	0	0	10	2	0	2	0	0	4	0	0	0	3	0	0	0	3
Diving figure	51	1	3	13	4	0	14	6	0	3	1	1	1	3	0	3	5	0	0	0	3
1764 grafiti	9	0	2	1	0	0	2	2	0	1	0	2	0	1	0	1	0	0	0	0	0
Eye of sun/ stick family	12	0	0	0	0	0	2	1	0	1	1	0	1	3	0	4	0	0	0	1	1
Owl family	4	0	0	0	0	0	0	0	0	0	0	0	0	0	2	2	0	0	0	0	0
Razz panel	19	6	0	0	0	0	7	1	0	1	1	0	2	0	0	0	2	0	1	0	1
Fish face	29	2	0	0	0	2	7	3	0	2	4	0	8	0	0	0	5	1	0	1	3
Comet-eclipse	23	0	1	6	1	0	4	2	2	0	0	1	3	0	0	0	4	0	0	1	2
Rock tower total	40	3	0	1	0	1	12	4	0	1	5	0	5	1	0	1	6	1	0	2	7
Serra do sol here i am	8	0	0	0	0	0	0	0	0	0	0	0	1	2	0	2	3	0	0	0	0
Serra do sol north solstice	25	0	0	0	0	0	16	5	1	1	2	0	0	1	0	0	8	0	0	0	0
Total paintings at panels	244	12	10	21	5	3	74	26	3	12	14	4	25	11	2	13	36	2	1	5	20

Fig. 5.20 Red and yellow handprints (nearly all are right hands) in the wall space between several large dots and circles. The yellow hand in the left corner was drawn, but the yellow streaks in the upper left corner are fingerprints dragged across the surface of the wall. Photo was Dstretch_lds yellow enhanced

The Orifice Mystery

In addition to the rock art panels that potentially track the sun (see next chapter), a few other panels contextualize the pictographic trail. Rock art motifs within panels remain diverse, not lending easy categorization to any specific trend. Also, there is no clear theory explaining one wall filled with art next to an adjacent wall that is devoid of any art, other than the theory of recording the sun ecliptic.

Wall smoothness does not appear to be the decisive factor. For example, the panel in Fig. 5.21 depicts some uneven segments of the wall, and the rayed concentric circle left of center in the panorama was painted on the most protruding segment of the wall. It is also angled a bit upward to the sky, and the photo shows that it is bathed in sunlight even as the wall above it is in shadow. There are also arcs and small dots painted near two altered orifices. A humanoid face next to the circular orifice appears to have its tongue out, and a zoomorph resembling a fish is to its left. The meaning behind this art is lost to me, but beneath the circular orifice is an orifice with a straight horizontal roof. So, orifices have been altered in diverse ways perhaps to reflect diverse reasons.

Fig. 5.21 Gigapan panorama of the "Eye of Sun/Razz" panel (top) displaying the rayed concentric circle on a protruding surface in the sunlight. On the right side of the panel is a circular orifice and a face. When zoomed in (bottom), the orifice appears to have been rounded and the face seems to "razz" an onlooker with its tongue stuck out. A zoomorphic fish is to the right of it

Another exeample of orifices altered in diverse ways includes painted orifices. One small circular orifice has its rim painted red, and two others next to it may have originally had the same treatment (Fig. 5.22). There are similar painted orifices at several rock art sites on *Serra da Paituna,* including at *Painel do Pilão*. These orifices are typically much smaller, about the size of a finger, like the one depicted here. Because the red rim forms a circle, they have been suggested to be related to the sun—specifically as places where sunlight might illuminate the interior. In this image, the orifices lie just to the left of the sun/shadow line, though the shadow might be cast by the bar hanging above them. The bar itself may have been intentionally altered into this shape, potentially reducing the sunlight exposure to the orifices, limiting their illumination to fewer days. However, this specific feature was not studied or photographed with that hypothesis in mind.

There are several altered orifices throughout *Serra da Lua*, and except for a concentric circle painted inside the orifice at the rock tower (see next chapter), the rest are quite mysterious. Perhaps some served as caches to store tools or food, but no such clues have been found to substantiate those or any theories. One orifice has a disembodied face drawn in red paint inside it (Fig. 5.23). This hole seems too small

Fig. 5.22 Red circumscribed orifice at *Serra da Lua*

Fig. 5.23 Small orifice with disembodied face inside (two red dots for eyes and downward arc forms a frown). The hole is ~12 cm wide and the "face" is ~8 cm. It faces west, between 272° and 290°, but exact cardinal direction is unknown

Fig. 5.24 Gigapan panorama of the "Fish Face" panel (left). There are many orifices on the wall, including the two larger holes in the wall at the left edge, and several extending linearly beneath a horizontal ridge higher from the natural ground as one walks right (bottom right the author is depicted). The wall runs east-west, facing north, as *Serra da Maxira* can be seen to the west (top right)

to be practical, however, and it is uncertain whether some edges are natural or intentionally altered.

The mystery of orifices is further demonstrated with this panel (Fig. 5.24). There is no dominant theme, other than perhaps the lack of sun-related motifs, presumably because it does not face the path of the ecliptic and receives no direct sunlight during the day. It does, however, contain several zigzags clustered in the top left corner, more zoomorphic motifs than other panels, two detailed disembodied eyes, and numerous orifices arranged linearly along a horizontal ridge—possibly cut or altered in the wall. Some of these orifices appear intentionally altered, while others seem natural, as some are half or fully circular. It is possible that some holes were bored completely through the wall. The ground dips significantly lower to the west, placing some orifices, and the two red circle eyes painted above them, at least 3–4 m above the trail. This suggests that either the ground was much higher in the past and has since eroded, or that scaffolding was used to reach such heights. The orifices might have been altered for storage, or they could have held no functional significance at all. I did not examine the two larger holes on the left side, but investigating them in the future to determine if they contain artifacts would be worthwhile.

I call this panel "fish face" due to the disembodied face with crescent ears located high at the far left of the wall. The crescents would be fins on a fish, though, not ears. To the right of this face is a possible zoomorph resembling a frog, although it could alternatively be a humanoid. It has three digits at the end of its fore and hindlimbs, a square torso with a circle in the middle, and a pentagonal head with no neck, so it seems more likely an animal. However, between its hindlimbs are either male genitalia or a double-flanged tail. Perhaps it is male, whether animal or human. Two pairs of eyes appear to the right, followed by a partial figure-8 with a hook on its top right side. This partial figure-8 resembles a pictograph on the ceiling at *Gruta Itatupaoca* (a cave on the south flank of Ererê) and resembles a pictograph at the base of *Pedra do Pilão* on Paituna (Fig. 5.25). I suspect this may be another stellar asterism because the symbol is positioned high and facing south at Itatupaoca and north here. If so, it suggests the entire panel depicts stars and constellations in the northern night sky. The one at Itatupaoca is on the ceiling and outlined in black. The one at *Pedra do Pilão* faces the south and has its hook facing down.

The disembodied face above the hooked partial figure-8 image has two concentric circles for eyes and a horizontal bar for a mouth. Dstretch image enhancements more clearly display them, and two more drawings even higher that look like a pair of eyes and a pair of distorted figure-8s (or birds) perched together (Fig. 5.26 left). A zigzag motif is visible on the right side of the photo. It could reflect clouds, water, lightning, or something else entirely unknown, but there are a total of three zigzags, all positioned high on the left side of the wall. Below those images, beneath a circular orifice, is a faint drawing resembling a concentric circle in the regular photo, but the enhanced photo (Fig. 5.26 right) reveals the center of a possibly petaled flower. Several more drawings also appear in a smeared jumble. In the right corner of the photo, there appears to be the profile of a raptorial bird, but the wall is chipped off

Fig. 5.25 Partial figure-8 with hook facing north at *Serra da Lua* (left photo). Colored- in red pictograph with blue or black outline and accents on ceiling at entrance to *Gruta Itatupaoca* (center), taken October 30, 2009. It is small and faint, but often associated with a frog, despite a depicting a possible tail. A lizard is more likely, and they sunbathe on the rocks of Ererê. Similar red pictograph shape with tail at base of *Pedra do Pilão* on *Serra da Paituna* (right photo), taken December 8, 2008, faces 216° south

Fig. 5.26 Image enhancements of pictographs at "fish face" panel. The Dstretch_lbl enhancement of black colors to contrast the paint in the shadows reveals more paintings on the ledge above a disembodied face with concentric dots for eyes (left). A faint painting resembling a concentric circle in the regular photo is revealed as possibly the center of a petaled flower design in the Dstretch_lrd image (right), which enhances red color

Fig. 5.27 Yellow humanoid motif where a portion of the wall was removed near the southern beginning of the pictographic rock art trail

where the head would have been drawn. To the right of the bird's foot is one of the single eye pictographs mentioned earlier. The eye is a half oval with a vertical oval-shaped eyeball. Overall, this panel has numerous eyes: all are unframed, two are singular, and many are paired.

Perhaps the numerous eyes are meant as an analogy for the numerous orifices. The mix of images found in nature (flowers, fish, and birds) and the numerous disembodied eyes and faces suggest that the panel might refer either to real ecological resources in the landscape, or to stellar symbols of ecological topics.

Although not a true orifice, this alteration to the wall (Fig. 5.27) is just as mysterious. A section of the wall was chiseled out based on the numerous flake marks along the edges. However, the concave portion with the yellow humanoid pictograph painted on it seems quite smooth. Perhaps upon chiseling and loosening the edges, a chunk of the wall broke and fell in a symmetrically concave fracture due to natural impurities or sedimentation layers. The rarity of such a fracture could have been enough to mark it permanently with rock art. We might never know for sure, which adds greater mystery to some of the alterations made to the walls at Monte Alegre.

References

Manning, C. D. and H. Schütze 1999. Foundations of Statistical Natural Language Processing. MIT Press, Cambridge, MA.

Pereira, E. 2002. *Arte Rupestre Na Amazonia Para*. Belem: Museu Paraense Emilio Goeldi.

Silveira, L. T., R. V. L. Pinheiro, and S. V. L. Pinheiro. 1984. *Roteiro Espeleologico Das Serras Do Ererê E Paituna*. Belém: Inedito.

Shannon, C. E. and W. Weaver 1949. The Mathematical Theory of Communication. University of Illinois Press, Urbana, Illinois.

Chapter 6
Archaeoastronomy at Serra Da Lua and Serra Do Sol

Sky Records of the Sun

I first recognized archaeoastronomy alignments with the rock paintings at *Serra da Lua* during the 2008 field season. The first clue was the fact that all the known rock art on *Ererê* faced west, including the rock art at Mirante despite the site being on the eastern end of the hill range. Another major clue was the fact that, even among the angled cliff walls of *Serra da Lua,* some perfectly smooth walls had no paint despite being next to a wall covered with several images. The prevailing winds and rain blowing from east to west could have been a reason, perhaps having eroded any eastern-facing art long ago. However, there was eastern-facing art on *Paituna* hill, plus rain still sometimes reaches the painted walls of Serra da Lua. Additionally, there was no rock art on the southwest cliffs of *Ererê* even though they too would have been less exposed to erosion, so the orientation of paintings exclusively facing west is unique to *Ererê.*

To confirm this observation, I began recording the exact cardinal directions that the paintings faced. The data revealed that the *Serra da Lua* rock art sequence begins along the hillside wall at a point near the southern winter solstice, and that a concentric dot perched on a pedestal was one of the first paintings in the sequence. It seemed to depict the exact horizon it was facing when the sun's path sets at the time of the solstice. To test this, I was able to return to the site on the 21st of December 2008 (and again in 2010 and 2016) to observe the winter solstice. Since then, I have found that all the concentric dot or concentric circle images that span *Serra da Lua* and *Serra do Sol* align with the path of the setting sun (or moon) between the southernmost and northernmost sunsets—corresponding to the summer and winter solstices. To avoid confusion with calendar dates, and the fact that the research area technically falls in the southern hemisphere (which reverses the terms "winter" and "summer"), I refer to the solstices of long ago as "southern" or "northern" solstices

C. S. Davis, *Before Sunset*, Conflict, Environment, and Social Complexity,
https://doi.org/10.1007/978-3-031-93373-8_6

instead of winter and summer solstices. This more specifically refers to the sun's relative position on the horizon as it sets.

Recognizing the passage of the seasons is perhaps something we take for granted in higher latitudes because we have so many other obvious signs. The northern latitudes, for example, experience vast differences between summer and winter seasons, the most obvious being the change in daylight hours, the shedding of deciduous tree leaves, the cold temperatures, and even the precipitation of snow. However, tropical environments—like Monte Alegre's location at about 2° south latitude—do not have any of these signs. The passage of the seasons—i.e., a solar year—is only truly recognizable by the migration of the sun between southern and northern solstices. The only problem with this method for a tropical rainforest environment, however, is that it requires clear vision of the horizon. The elevation that the hills of *Monte Alegre* provide satisfies this requirement.

Another peculiar observation that possibly inspired early artists are the sun and moon halos (Fig. 6.1), which, due to the intertropical convergence zone (ITCZ), and local evaporative and orographic effects, seems to occur here more often than many places in the world. The halos are the best explanation I can offer for the high frequency of concentric circles and dot paintings. It seems apparent that the early artists began to pay close attention to the movement of the sun, and maybe the moon too. Once they began making images of the sun during different times of the year, perhaps when it formed sun halos, they were well on their way to recognizing and establishing a record of a seasonal solar cycle.

Today, solstices occur on the 21st of December and June, and I took observations and photos of the sun and the paintings directly on those days in 2010 and 2011. Today, the December (southern) solstice at *Serra da Lua* sets on the horizon at 246.4°, and the June (northern) solstice sets at 293.4°. Not all paintings have ideal precision that satisfies all skepticism, but a slight lack of precision should perhaps be expected from the ancient artists who, painting by hand on narrow ledges, were using naked eye observations (or gnomons) for precision, and the sun's luminous intensity would also have been difficult to gaze directly.

Fig. 6.1 Sun halo photo (left) near Serra da Paituna, May 2011. Moon halo (right), photo taken January 29, 2010

Assuming the ancient artists at *Serra da Lua* began painting around the same time as indicated from radiocarbon dates at *Pedra Pintada* and *Painel do Pilão*, they would have experienced solstices over 13,000 years ago on the 6th of March and the 6th of September, according to StarryNight software. Due to precession (the wobble of the earth's axis around the polar stars), the months that the solstices occurred were far different than today. Although this may have been significant for Milankovitch climate variations, the sun was setting on the horizon only ~0.8° further south and north than it does today due to obliquity (the tilt of the earth's axis). According to the IAU equation (Capitaine, et al. 2003), obliquity (ϵA in arcseconds) $= \epsilon 0 – 46''0.84024t - 0''0.00059t^2 + 0''0.001813t^3$, where t is measured in Julian centuries of 36,525 days since Jan 1 2000 (or in our case, before 2000, so t is negative). The value $\epsilon 0 = 23.44°$ today, which is 84381.448″ (arcseconds). Calculating for 13,000 years ago, earths axial tilt was ~24.2°. The southern solstice, therefore, occurred on March 6, 11,200 B.C., at ~247.2°, and the northern solstice occurred on September 6, at ~294.2° northwest on the horizon.

From the *Serra da Lua* rock art sequence about 72 m up *Ererê* hill, the edge of *Serra da Maxira* hill blocks direct sighting of the horizon at level ground. However, the painting on the cliff wall depicts the sun above a "pedestal" anyway, not the sun above a horizontal line. This pedestal painting faces ~248° southwest, correcting for true north and it appears to depict the sun setting on top of a small rocky perch on the hill of *Maxirá* in the horizon. The "pedestal" that the sun "perches" on is not clear today but may have been the edge of the hill to the left 13,000 years ago when the sun set ~0.8° further south (Fig. 6.2). Another important clue is that the painted pedestal has a yellow "glow" on its side, the same side that the sun would have glowed around the edge of *Maxirá* 13,000 years ago when it set further to the left in the photo.

Also, we see another painting of a "smiling" rayed concentric circle—that I also believe represents the sun—just a few centimeters above and facing further south at 242° southwest, making it the circle image furthest south. This smiling sun is positioned higher on the wall, perhaps aligning to the sun when it is higher in the sky. Also, the hill itself is at an altitude of approximately 126 m above sea level. Starry Night software reveals that when viewing the sun, a bit earlier in the day, it appears further south than at sunset on the horizon. Verifying with an online database (SunEarthTools.com 2024), on the day of the solstice, the sun does indeed swing out further south before descending a bit backward to "set" (Fig. 6.3), as indicated by the paintings (Fig. 6.4). It reaches an azimuth of 242° on the day of the solstice at about 3:00 PM at *Serra da Lua*. Also, *Maxira* Hill has no doubt eroded since the paintings were first made. And the sun sets ~0.8° to the right of where it set 13,000 years ago (which is about the length of your pinky finger held at arm's length). In fact, the sun's apparent diameter (of just the disc) is ~0.5°, so the sun was a little over a sun's length and a half further to the south (left) than today. When standing with your back to the painting and facing the sun as it sets, one gets the unmistakable feeling that this is exactly the phenomenon depicted.

The corresponding painting at *Serra do Sol* provides a more direct sight to the horizon. The rock art sequence at *Serra do Sol* terminates with a panel containing a

Fig. 6.2 Southern solstice pedestal pictograph at *Serra da Lua* (above left), and with D_stretch-lds enhanced yellow color (above right). Photo of the December 21st solstice sunset in 2008 (below) with arrow to the left showing sun's path ~13,000 years ago when earth's tilt was ~1° wider than today. Enhanced photo was taken a month earlier in November 2008

large concentric dot correlated with the northern solstice sunset (Fig. 6.5). Here we find the largest painted image in all of Monte Alegre, at ~1.3 m in diameter. Correcting for true north, the image faces ~294° northwest, which is closer to the position of the sunset on the day of the northern (summer) solstice 13,000 years ago (correcting for precession). Correctly so, the present-day sunset at ~293° was ever-so-slightly misaligned to this painting. However, one should note that the *Serra do Sol* foot trail is about 160 m above sea level at the concentric circle, and another 3 m higher on the wall than the trail. Here as well, the sun's intensity, higher hill elevation, and position on the cliff wall are reasons not to expect exact precision to the horizon, yet it closely aligns.

A closer inspection of the Serra do Sol concentric pictograph provides one additional clue to the manner in which the ancient artists may have aligned the painting to the sunset. A small circular depression is at the center of the image. Not only could this hole indicate the point by which the shadow of a gnomon could have been

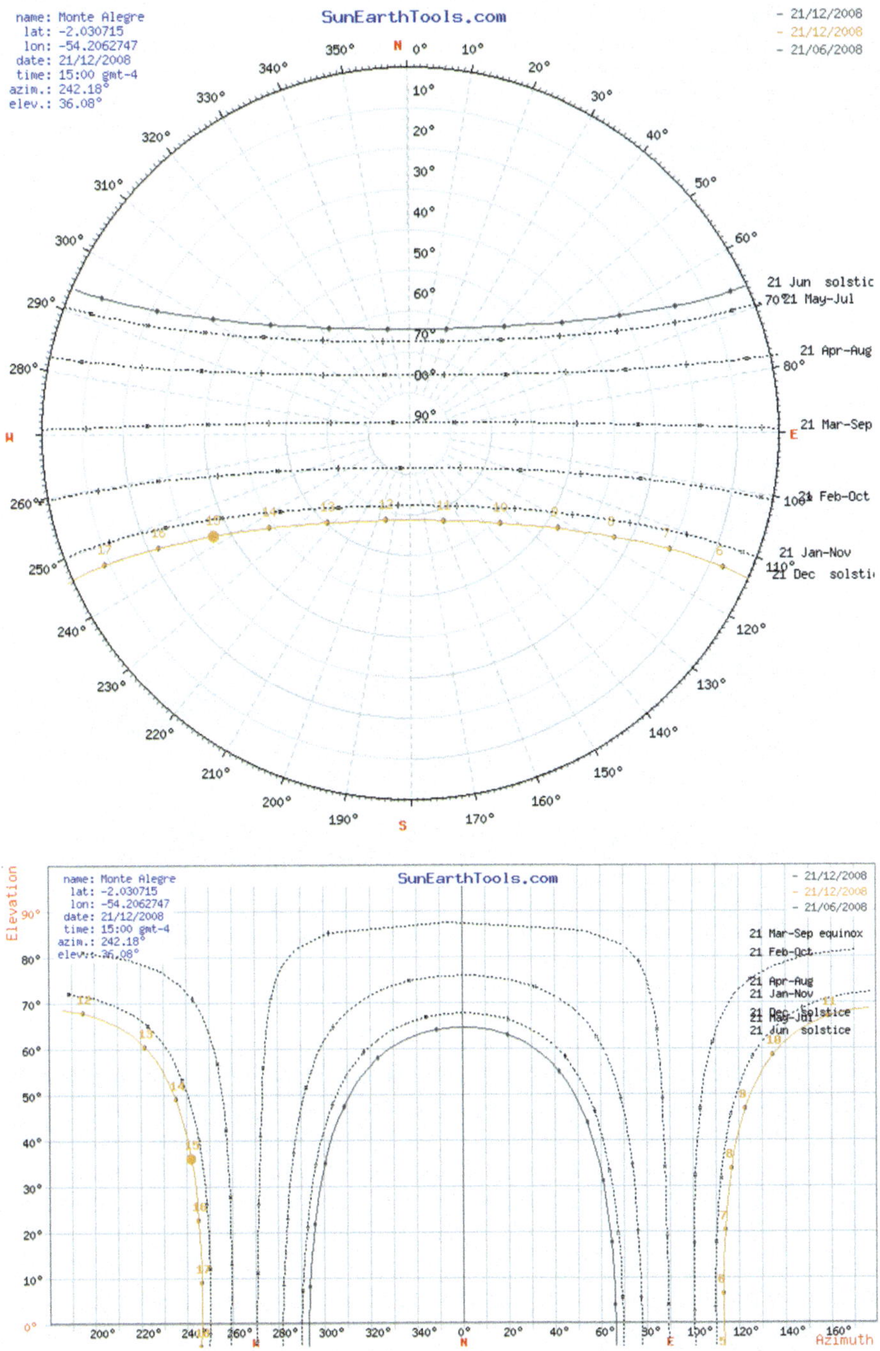

Fig. 6.3 Sun Chart of December 21st, 2008, at Serra da Lua, 3:00 PM (15:00) (not adjusting for daylight savings time) displaying the local azimuth of the sun at 242°. Further evidence that the sun appears to bow further south before descending north, just like the pictographs depicted on the cliff wall

Fig. 6.4 Sun emerging over the crest of *Ererê* hill further south than its position later at sunset. This photograph was taken on December 21, 2016, at 12:59 PM

cast (so that the artist(s) didn't have to gaze directly at the sun's luminous intensity), it also may have served as an axis to circumscribe the very symmetrical concentric dots. Perhaps a string was attached to a stick positioned in the hole and swiveled around to outline the red and yellow boundaries. During my efforts to take a photo of the pictograph at parallel camera angle to the wall, I attached my camera to a pole and used a delayed shutter to take the picture. Several attempts captured the shadow of the pole in the image (Fig. 6.6), which was distressing in the moment, but in retrospect provides an example of how a gnomon's shadow could make sunset recordings more precise.

There is also a concentric painting facing the approximate position of the sun on the equinox (Fig. 6.7). A third piece of evidence for the solstice hypothesis comes

Fig. 6.5 The largest concentric dot painting in Monte Alegre in this photo at *Serra do Sol* (above left), and a photo of northern summer solstice sunset in front of the pictograph (above right) on June 21, 2011. The entire Northern Solstice panel in front of the solstice sunset is shown as revealed with D_stretch-lds (below) enhancement to emphasize all colors, but yellow the most, in this Gigapan panorama of stitched photos

from large concentric circle paintings at a location in front of the equinox sunset. Perhaps the more practical significance of this location is that it coincides with the sun reemerging from behind the *Maxirá* hill. However, that apparently coincides with the position of the setting sun on the equinoxes. The sun begins to set behind the shadow of the hill as the calendar nears the southern solstice, and it clears the hill 6 months later, on its daily movement across the sky to the northern solstice.

The equinox concentric circle painting is not as large or preserved as the solstice images, and a current tree sapling made photographing it more difficult. However, it is right next to another concentric circle painting, which is the only one with a portion missing from its central dot. It could have been a portion of the wall that chipped, but paint beyond the chip boundary suggests the "hole" in the dot was

Fig. 6.6 Gnomon Shadow. The shadow is of my camera tied to a stick when I attempted to take a photo of the concentric circle perpendicular to the wall, but it is about 2.73 m (~9 ft) above the trail today. Note the small circular depression near the center of the red dot. It may indicate how the painting was made so circular. Note the much smaller concentric dot to the right may be further evidence that the sun's progression was "tracked" days before the solstice

intentional. Perhaps it represents the moon? It less likely represents a partial eclipse, since sun intensity would have still made such a distinction very difficult with unaided observations. The two close concentric paintings could alternatively represent the artist(s) attempts at precision for the equinox, since mostly just the counting of days between solstices, followed by those days being divided by 2 could "discover" the equinox sunset. There is another way within the tropics to determine the equinox, though. It is the day when a gnomon does not cast a shadow at high noon (Fabian 2001; Reichel-Dolmatoff 1982). Such a day depends on the day that the sun passes directly through the zenith of the sky. For Monte Alegre at 2° south of the equator, that would not occur on the day of the equinox, but instead a few days before (for Spring) or after (for Autumn). This concentric circle "hole" is positioned approximately where such an event would occur here, if indeed this rock art trail represents the days of the year.

The effects observed while walking along the pictographic trail is important to conceptualize. An observer standing on the foot trail with his back to the sun pedestal painting (Southern Solstice) would see the sunset on the day of the equinox out

Fig. 6.7 Equinox sunset circumscribed dot (center) with Dstretch_lab photo enhancement to emphasize white and black colors. To the right is part of a red circumscribed dot with a "hole" in its central dot

of the right corner of their eye (see Fig. 6.8 Middle). Additionally, from that same position, they would not be able to see the sunset on the northern solstice at all. Instead, they would need to walk along the initial curve of the wall. Even if an observer were standing underneath the equinox painting, they would see the sun set far to their right side on the day of the northern solstice (see Fig. 6.8 Bottom). Similarly, an observer with his back to the large northern solstice pictograph at *Serra do Sol* would not be able to see the sun set on the day of the southern solstice at all, due to the angle of the local wall there (Fig. 6.9).

Using *Serra da Maxira* as a horizon marker, the ancient artists might have walked along a foot trail that follows the natural curvature of the hill from southwest to northwest (Fig. 6.10). Four other concentric dot images along the *Serra da Lua* trail, and several other rayed circle images from *Serra da Lua* to *Serra do Sol* all face positions between the two solstices (Fig. 6.11). Closer scrutiny of their cardinal orientations suggests there are at least two different sets of solstice ranges (Fig. 6.12). Each of these segments, and/or the entire segment from *Serra da Lua* to *Serra do Sol* might therefore provide a walking tour of the solar year, and perhaps some of the resources available at specific times. The various paintings depicting animals, faces, figures, and designs might be stellar asterisms or constellations that the ancient sky-watchers envisioned in the night sky, further dividing the intervals between solstices and equinoxes. Our own 12 zodiac constellations are the stellar correlates to the solar year being divided among 12 astronomical intervals determinable at night. Perhaps the ancient Paleoindian artist became inspired to "annotate" their solar calendar by using paintings at *Serra da Lua*, as though it were a walking almanac.

Fig. 6.8 Migration of sunsets with *Serra da Maxira* as a horizon marker as viewed from the concentric circle pedestal pictograph during the December 21 solstice (top) and March 21 equinox (middle). From this position, the equinox does not align, but it to the right. Likewise, the June 21 solstice sunset is off to the right as photographed from beneath the "equinox" pictograph (bottom—a tree sapling is seen in the foreground)

The Diving Figure

Possibly the most intriguing panel at *Serra da Lua*, and arguably the most famous, is the pair of anthropomorphic concentric circle paintings located ~13 m from the beginning of the pictographic trail (Fig. 6.13). Both pictographs start at about 2 m

Fig. 6.9 Photo of landscape taken looking southeast from near *Serra do Sol*. Likewise, the December 21 solstice sunset is not visible at all from this location. In the distance (center of image) is the northern edge of *Serra da Maxira*, and the horizon where sunset would occur on the equinox. This photo demonstrates how the painting at *Serra do Sol* only aligns with the northern solstice sunset

Fig. 6.10 Photo of the western curve of *Ererê* hill taken from *Serra da Maxirá*. This view shows *Serra da Lua* (the white cliffs) in the center in the distance. South is to the right side of the photo, and north is left. *Serra do Sol* is located on one of the north cliffs

high on the wall from the current trail ground. The upside-down image is longer, with a total length from rayed head to tip of the leg of ~1.6 m long (the concentric circles have a diameter of ~0.55 m), while the upright image is about 0.95 m long

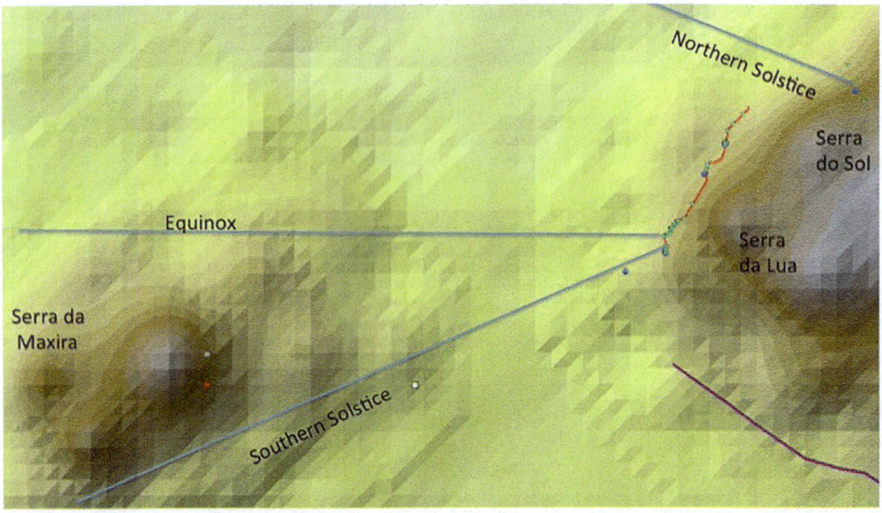

Fig. 6.11 Map of the *Serra da Lua*/*Serra do Sol* western horizon sunsets. The rock art trail (red line) and the main compass alignments to the horizon sunsets (blue lines). The red point on *Serra da Maxirá* is the location of the painted cave: *Caverna do Diabo*. Green points are panel paintings. Blue points are camp stations. All positions are to scale

from top to bottom (the concentric bands have a diameter of ~0.6 m). Both drawings are unique in the sense that they are drawn seemingly with bodies.

The larger image has two concentric circles around a small central dot, with rays radiating down from the outer circle, and the body is above the head. Two long spindly lines emerge downward too, perhaps representing arms (or smoke?), and a downward crescent is positioned like hips supporting two bars for legs above it. The image is only drawn in red (although the background wall appears to have been primed with a whitish paint) and the body is filled in with red color. The adjacent anthropomorphic figure is right side up (assuming the concentric dot is the head), with a more triangular body and yellow color filling in the interior dot that is circumscribed by a thick red circular band. This image appears to lack "arms." The diving figure faces ~259.5° west, while the upright figure faces ~252° west. Consulting the annual sun path online database (SunEarthTools.com 2024), the sun sets at about 252° on the western horizon by January 29 (Fig. 6.14), which is 38 days from the winter solstice. Although the diving figure is higher on the wall, the sun doesn't reach ~260° until about February 21 (Fig. 6.15), which is another 24 days after the first one (SunEarthTools.com 2024). However, the "diving" pictograph is positioned further to the south of the smaller image. Therefore, perhaps a different meaning can be taken from this unusual depiction.

The upright figure is perhaps the "Lua" (moon) of which the area gets its name, however, it could be a representation of the sun instead. Just like the concentric band at *Serra do Sol*, the concentric red and yellow bands of this pictograph have a central indentation too (Fig. 6.16). The red central dot covers much of the indentation.

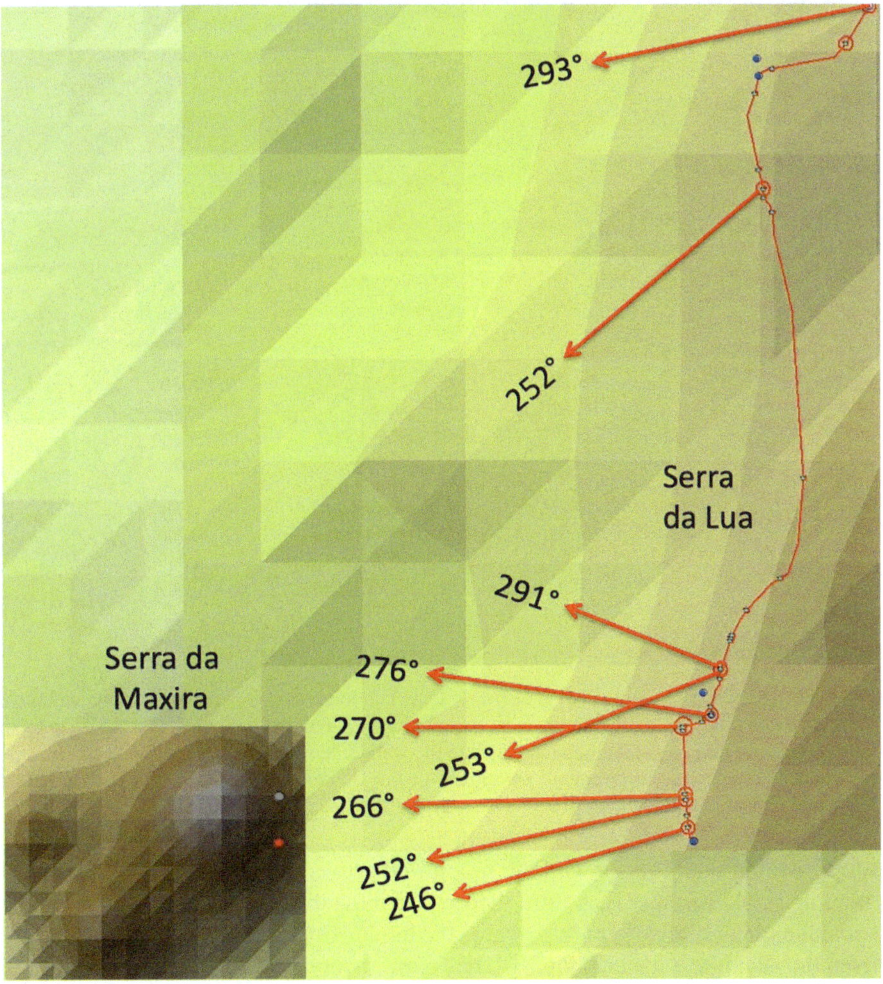

Fig. 6.12 Cardinal directions of the "sun" pictographs, whether concentric or rayed circles and dots. There is an initial set oriented between 246° and 291°, and then there are the further two sites that arrive at the rock tower, which cover 252° and 293°. This trail does not include the extended site of Serra do Sol, but a pictograph at the rock tower preserves a painting alignment with the northern solstice too

As discussed with the former, the indentation on the wall may have formed from some sort of central pivot from which a symmetrical circle was traced.

Numerous different theories can be proposed for the diving figure panel. I will briefly discuss four. One theory is that both images represent the sun, and the diving figure is the sun drawn at a higher position, more intense and cresting over the edge of the hill, while the smaller image is the sun setting behind Maxirá hill later at dusk when it is less intense. A second theory is that the diving figure is the sun, and the standing figure is the moon visible during the daytime (or vice versa), and the artists

Fig. 6.13 The Diving Figure panel at *Serra da Lua*. The focus of this panel are the two anthropomorphic concentric circle pictographs next to each other. A red and yellow spot are next to them, to the right, and numerous paint handprints line the wall underneath them. The diving figure is the pictograph in the center of the photo positioned upside down. Four theories are discussed about this panel, possibly representing observations seen in front of it

were tracking both the sun and the moon to discover how or when they pass each other in the sky (eclipse?). A third theory is that the diving figure is a second sun … something incredibly bright that rivaled the luminosity of the sun while it descended from the sky, like a meteor impact. The fourth theory is similar to the third, except that it is a comet instead of a meteor.

Going through each theory in greater detail, theory 1 is that they both represent the sun, and what we see is the idea that the ancient artists imagined the hill coming alive when the sun appeared to rest on top of the hill (Fig. 6.17). This period (between equinox and winter solstice) could have been given cultural significance as the dry period—as it coincides today—or that *Caverna do Diabo* on *Maxirá* hill is an underworld entryway for the sun to travel at night (which could explain why that cave is hot). Indeed, there appears to be a painting of a turtle on a ground slab deeper in that cave, with the turtle being a common representative of the sun among Amazonian cultures (Davis 2023).

The second theory about the two representing the sun and moon has supporting evidence by the two large red and yellow dots to the right of the diving figure in the photo. The red dot has an extended "hand" holding the yellow dot (Fig. 6.18). The red dot also has a yellow arc around its circumference to the left, and the squiggly

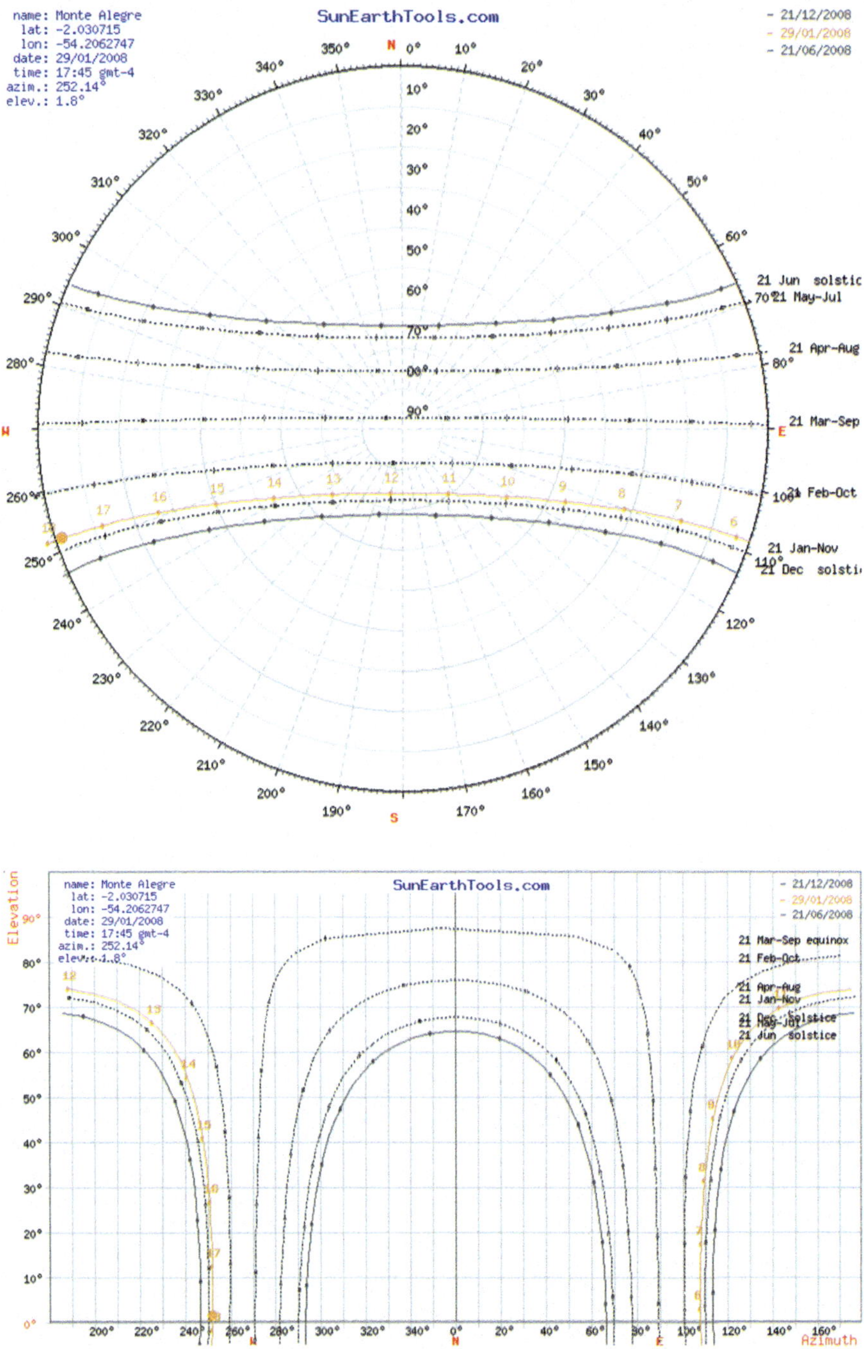

Fig. 6.14 Sun Chart at Serra da Lua on the day that the sun sets at ~252°. Today that corresponds to January 29 at ~5:45 PM local time (not accounting for daylight savings time) (top left data)

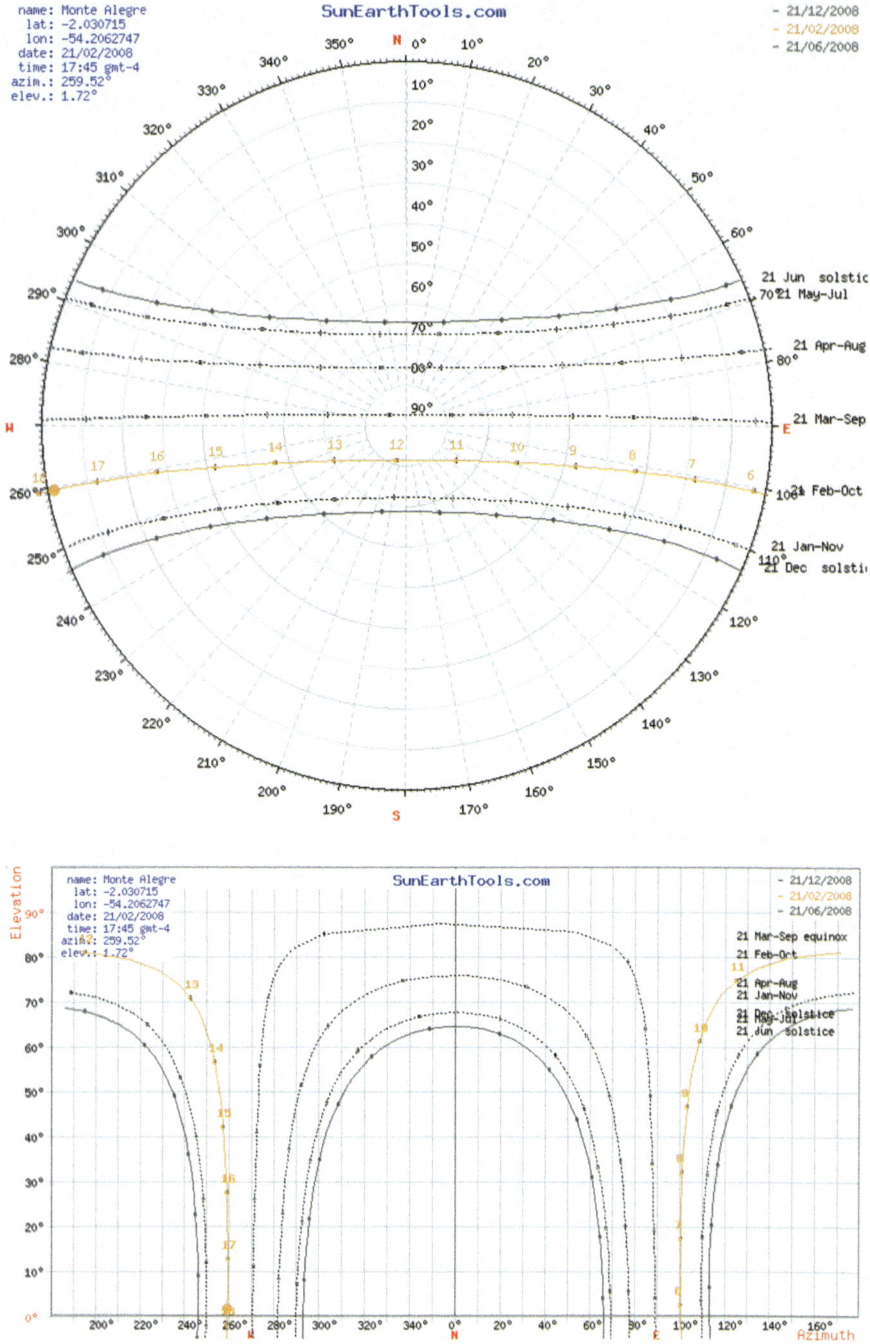

Fig. 6.15 Sun Chart at Serra da Lua on the day that the sun sets at ~260°. Today that corresponds to February 21 at ~5:45 PM local time (not accounting for daylight savings time) (top left data)

Fig. 6.16 Concentric red and yellow bands of the upright figure at Serra da Lua exhibits a central indentation in the wall, from which a central pendulum or implement may have pivoted around to trace symmetrical circles. The diameter is ~0.6 m

line of the diving figure seems to extend to touch the red spot. This makes at least three of the circular figures directly connected. Their connection could, therefore indicate that they are the same object observed at different days apart to track their movement (slippage) north or south each day. Or perhaps the artists are painting the moon eclipsing the sun. In this scenario, perhaps the yellow dot represents the moon, and the red dot represents the sun. The personified image with the red and yellow concentric dot could therefore be a solar eclipse with the sun's corona circumscribing what the artist(s) recognized to be the moon.

The third theory is that the diving figure is a bolide meteor that entered the atmosphere and flared up to be as bright (or brighter) than the sun. There are many challenges to this theory, or any rapid and unpredicted event, however. Any meteor entering the atmosphere would have passed quickly (seconds to minutes), and it must have been far away (to not damage the immediate area or inhabitants like in Tunguska, Russia on June 30 A.D. 1908—which was just about on the day of the summer solstice). If such a sudden and rapid event were the case, the artist(s) must have come to the cliff sometime later to draw a record of it. This doesn't disqualify the theory since the same could be said about an eclipse (in which total blackout lasts only a few minutes), but it seems less likely if the event itself occurred rapidly.

The fourth theory—a comet around the end of the Ice Age—does bring to mind the comet impact hypothesis, which has been proposed to cause the onset and/or end of the Younger Dryas. Comets appear several times in the rock art at *Serra da*

Fig. 6.17 The diving and personified figures might represent the sun (and/or moon) descending above *Maxirá* in the afternoon sky (top—photo taken on November 8, 2009, ~5:51 PM). A more dramatic event if a sun halo (bottom left—recreation) or moon halo (bottom right—recreation) were observed. The paintings could be animistic representations of the sun or moon diving from the sky to "walk" upon the earth, the period of the year which corresponds to the dry season now, when travel on foot is easier

Lua and other rock art sites in *Monte Alegre*. So, if the diving figure is a comet, perhaps it is an indication of an incredibly bright comet that was visible during the day. The orientation of the drawing would suggest that the object moved across the sky from east to west (~260° west in the same path of the sun) during the late afternoon as the sun (the upright drawing) was setting. The time of year, as stated earlier, would be potentially 38–62 days after—or before—the winter solstice. Although the months in which they occurred were vastly different from today, 38–62 days before the winter solstice of December 21st today would give us a time frame between October 20 and November 13, a period coinciding with the Southern and Northern Taurid meteor showers, respectively, today. It's just one theory, but if the comet impact hypothesis holds up to scrutiny, this painting may be a clue to such an event. Not on the day of the comet entering the atmosphere and impacting specifically, but perhaps weeks to days leading up to an incoming comet that began to shine as bright as the sun and appear during the daytime.

Fig. 6.18 Large red and yellow pair south of the "diving figure" at *Serra da Lua* enhanced with Dstretch_yrd, which emphasizes red color. The red marks beneath and within the yellow dot are now more visible. A red arc also extends from the red dot onto the yellow dot, and a partial yellow arc is around the left of the red edge. These spots were either earlier attempts at the symmetrical concentric circles, or they track sunset sightings further left or right each day

The Comet-Eclipse Panel

Another very intriguing panel at *Serra da Lua* is the one located just 12.5 m south from the rock tower. It appears to depict a bright star, a spoked concentric dot encroached on both sides, and an anthropomorphized comet (Fig. 6.19). With the top of the comet face being 4.46 m above the closest walkable trail, the rest of the pictograph higher on the wall is angled upward to the sky, and the panel faces a local azimuth between 302° and 309° northwest. Maintaining consistency in motif interpretation, I categorized the concentric dot as a representation of the sun. The encroaching arcs around it, its comparably small size, and the spokes may represent the diminishment of its luminosity to the point that it is only as bright as the star or planet above it. This therefore suggested to me that the painting was a possible depiction of a solar eclipse. The ray beneath it might represent the sun reemerge from the eclipse, since this is the direction that the sun's light would have reemerged. Interpreting the streaking face as a comet was a bit more problematic because the "head" of the comet painting was right side up in relation to the concentric circle, and in relation to the horizon that the panel faces. Comets do not usually become visible until after the sun goes down, and because their tails point away from the sun

Fig. 6.19 The Comet-Eclipse panel at *Serra da Lua* just 12.5 m away from the rock tower. The natural-color photo (above) and a sepia tone of the top portion (below) to reveal the faint handprints

(usually), the comet's head should be pointing down as well. The other two concentric dots to the left of the "eclipse" might be other planets.

However, I stumbled upon one possible answer to the dilemma of the comet facing upright instead of upside down after my radiocarbon dates came back with a

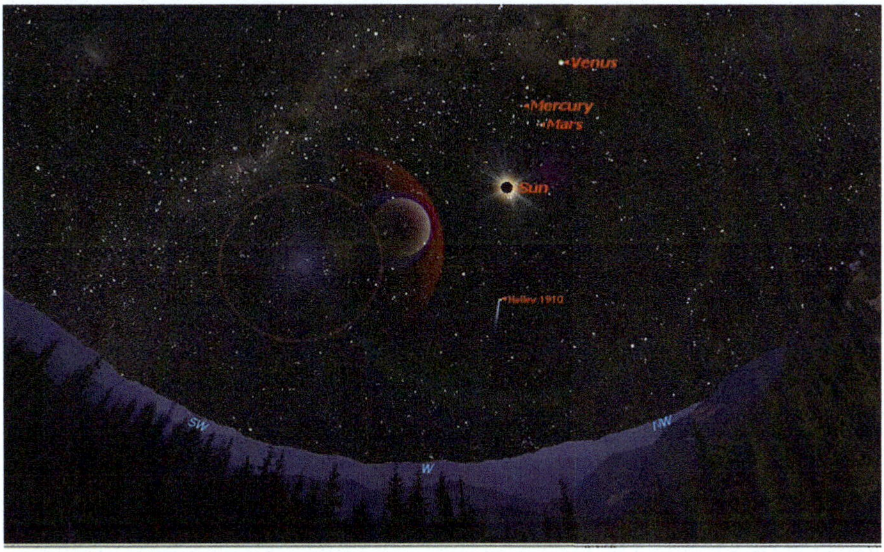

Fig. 6.20 A StarryNight™ recreation of the sky at *Serra da Lua* during a solar eclipse between 1:30 PM and 2:30 PM on July 28, 11,027 B.C.. The total eclipse may have revealed a comet that otherwise would have been invisible from daylight illumination

date range in which to search. I used StarryNight to search for potential solar eclipses around 13,000 years ago and discovered that one eclipse coincided with another rare event—the appearance of a comet! The comet is Halley's, which produces a serious challenge to astronomers because the oldest confirmed reference to visible sightings of Halley's comet was by the Chinese in 240 B.C. (Kelley and Milone 2011, pp. 131, 134), perhaps the Greeks in 466 B. C. (Graham and Hintz 2010), and possibly even earlier by the Chinese in 1059 B.C. (Pankenier 2013:216). But what StarryNight™ confirms is that if its present trajectory was maintained 13,000 years ago, it would have become visible during the day on the 28th of July between 1:30 and 2:30 PM because there was a solar eclipse, temporarily darkening the sky to reveal the comet right side up, below the eclipse (Fig. 6.20). The eclipse occurred in the sky at a local azimuth at about 302° and the comet was visible at about 288°.

Additionally, during the total solar eclipse, the planet Venus should have become visible just above it, just as a painting of a star is positioned just above the encroached circle painting in the comet panel. Furthermore, Mercury and Mars are just beneath Venus and to the right of the eclipse, just as two concentric dots appear to the left of the encroached circle. The painting, therefore, seems to mirror an extremely rare event that occurred 13,000 years ago—potentially the oldest evidence for such a phenomenon in the world. Unlike the solstice sightings, a depiction of a rare event like a solar eclipse was probably painted from memory shortly after the event, instead of by directly painting what was observed. The ancient artists certainly would not have been ready, with paint in hand, to draw such an unexpected event.

As such, we should not expect alignment to be as precise, and yet there is an impressive match between painting, event, and alignment here.

An alternate hypothesis for the comet-eclipse panel is that the spoked concentric dot represents a nova or supernova. This hypothesis stems from a similar representation at Chaco Canyon that many scholars believe represents the Crab Nebula supernova. It was the second brightest supernova ever recorded, visible around July 4th of A.D. 1054 with an apparent magnitude of -5.75 ± 1.25 (Collins et al. 1999:877)—as bright as the crescent moon. It was visible in the constellation of Taurus for about 2 years until April of A.D. 1056, and its remnants left behind the Crab Nebula. Chinese reports claim the supernova had a slight iridescent yellow color and was only about as bright as Venus, suggesting that it only would have been visible from dusk until dawn (Needham 1981, p. 205). StarryNight claims that at its brightest, it was four times brighter than Venus, making it visible in the daytime for approximately 23 days.

Assuming the supernova's brightest light appeared during its initial onset, it would have appeared in the Monte Alegre sky during the daylight hours, rising at 68° azimuth, peaking at 67° altitude in the sky, and setting at 291° azimuth—about 2–3 h before the sun (verified by StarryNight™). Considering that it would have appeared to travel in the same path as the sun, and that it was about as bright as Venus, which at the time was about 3 h behind the sun, the supernova could have easily been mistaken for the planet, if the culture was not typically a night-viewing culture. Through the course of its 2-year illumination, the supernova would have remained setting at about 290° azimuth, which is the north solstice position at Monte Alegre. Any depiction of the supernova, if it exists, would be expected at *Serra do Sol*, or perhaps a different wall along *Serra da Lua* that was angled to that azimuth.

There is no reason to believe this rock art panel was a record of the Crab Nebula supernova at *Serra da Lua*, it is only presented here as a consideration that a different supernova—as yet unknown—is an alternative possibility. Currently, it is a far less likely theory for the comet-eclipse panel. I checked all the known supernova reports and none of them seem to align with this panel the way that the comet-eclipse event does.

Owl Mates

Although all concentric circle and dot motifs align within the ecliptic, not every pictograph along the *Serra da Lua* trail does. Some paintings and panels are on walls facing further north than 294°. One of the most prominent and distinct is a panel of figures that resemble an owl pair (one holding a handbag), a splayed zoomorphic figure, and a small aquatic figure. There are also three red spots overlaying the splayed figure, a circle in the triangle of both owls, and they are both also bordered with numerous red points (Fig. 6.21).

I believe this panel represents stellar asterisms, including the summer triangle, the wing of Pegasus, the circlet of Pisces and parts of Aquarius. Today, these

Fig. 6.21 Owl Mates panel aligned to ~317° northwest along the *Serra da Lua* pictographic trail. I believe this represents constellations including the summer triangle and the wing of Pegasus asterism in ecological association with the great horned owl mating season in January. The great horned owl is normally solitary except during mating season when male and female pairs exchanges "hoots" back and forth at dusk

asterisms are part of different constellations, including Cygnus, Aquila, and Lyra, which appear low in this portion of the western sky at Monte Alegre in January. It is also more common for some Amazonian cultures today to imagine and include the dark portions of the sky into their constellations too (d'Abbeville 1614, p. 319b; Fabian 2001, p. 60; Nimuendajú 1946, p. 233). We can't know how old that custom may be, other than to state it is something to consider. Around 13,000 years ago, this part of the night sky appeared low in the western sky facing about 317° in late September (Fig. 6.22). However, due to precession (as mentioned previously), late September was about 1 month after the Winter (southern) Solstice. The time frame was the seasonal equivalent to January today. These asterisms may have been imagined to partly compose the Owl Mates pictographic panel, which would have faced this portion of the sky ~13,000 years ago.

What is significant about January? If the panel depicts a pair of owls, the most probable owl species depicted for the lower Amazonian basin is the great horned owl. Their ear tufts taper down to a triangle between their eyes, which most closely resembles the "ears" and forehead triangle of the painted pair. An ecological association also occurs with January and the great horned owl. The owl is normally solitary, except during mating season when male and female pairs exchange "hoots" back and forth at dusk, acquire a nest, and both incubate the eggs. The mating sounds of the horned owl dominate twilight at that time of year. The splayed figure could represent some sort of lizard, but without further cultural insight, this is just speculation. The "fish" figure seems a likely portrayal due to its round body and upside-down, v-shaped lower limbs. However, it is a small painting compared to the

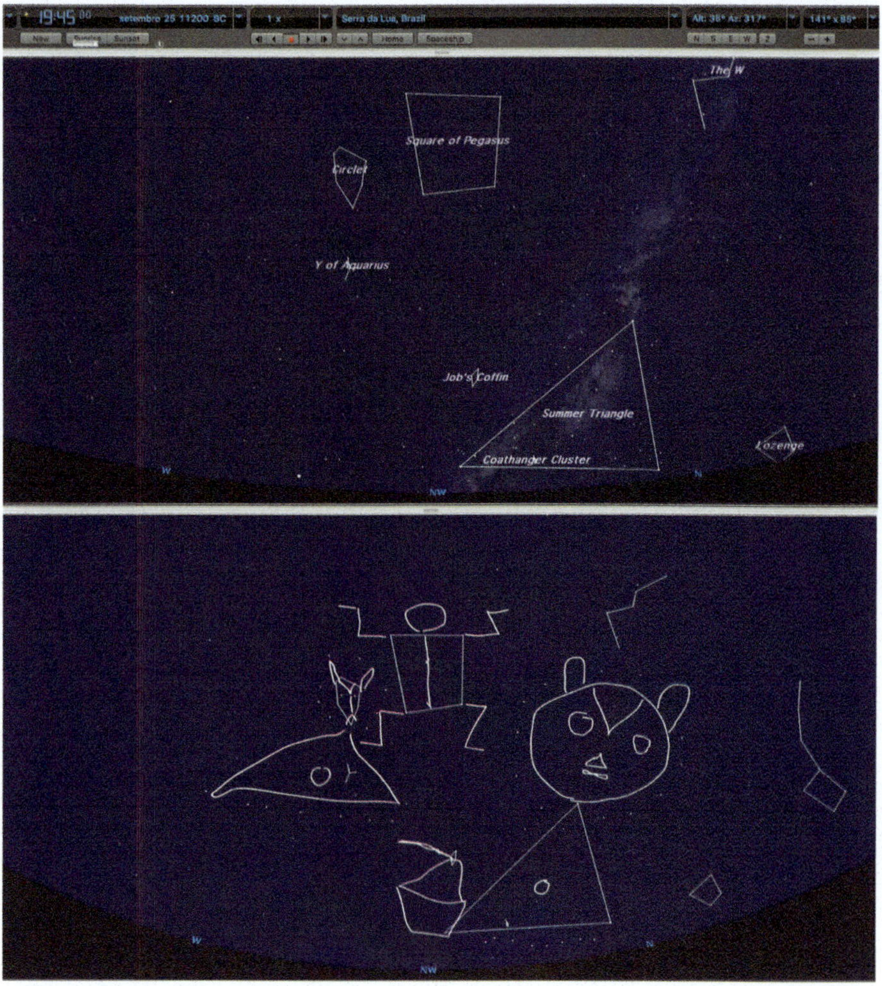

Fig. 6.22 StarryNight projection of the twilight sky at *Serra da Lua* on September 25, 11,199 B.C. at 7:45 PM facing ~317° northwest. The asterisms (top) from left to right include the circlet of Pisces, the Y of Aquarius, the square wing of Pegasus, the diamond of Delphinus (Job's Coffin), and the Summer Triangle, which lies with the band of the Milky Way. An imagined drawing of the "Owl Mates" is superimposed for comparison (bottom)

others, and any specific fish markings that would identify the species are unknown to this author. The handbag of the owl on the right side has the position that might correlate to the bright star Altair, and part of the constellation Delphinus in the night sky, if it represents a constellation. Ecologically, it could signify that this is the time of year to focus on gathering plants, seeds, nuts, and other dry-season foods.

Other theories were considered besides the main figures being owls—including that they may represent people wearing masks. If they are people, their triangular torsos suggest they represent females wearing dresses. However, such associations are modern and probably of incorrect cultural context for Stone Age Amazonians.

Fig. 6.23 The Rock Tower at *Serra da Lua* from its south-facing side November 16, 2007, photographed at 7:38 AM (left), at 10:14 AM angled more from behind the structure (center), and at 12:35 AM (right)

The idea that the faces are masks is always a difficult theory to disprove, but the numerous points around the faces and triangle bodies are not addressed by the mask theory and would therefore require a second theory. Even if they are masks, they seem to best represent the horned owl. The horned owl is reported to eat lizards and fish, so the other paintings are still possible correlations. As masks, however, the triangles would therefore be costumes, yet no arms or legs are depicted.

Although the mask theory is not disproven, the theory that this panel represents constellations is still best supported by current observations and knowledge.

The Rock Tower

The main rock art trail at *Serra da Lua* terminates at the rock tower. It is a free-standing structure ~10 m high, wider at the top, and the protruding portion extends over the base in the northwest direction. The trail approaches the tower from the south and terminates at its eastern base. However, moving off the trail to the west reveals the top portion starting to resemble a bird's head with a protruding beak. Although the tower is a natural formation, it is uncertain if part of it naturally eroded to resemble a beak or if the shape was intentional (Fig. 6.23). Rock art is visible near the top and at the base, but not on the north side. There are rayed and concentric circles facing the west, but the rock tower is the one place that also has a concentric dot and other images facing southeast (Fig. 6.24). The pictographs facing southeast are positioned high on the tower, even though the east-facing base of the tower is where the ground is highest—the direction from which one arrives to the tower from the trail. Full examination of the tower reveals that the east paintings are only positioned where the sun's direct rays strike it, when the sun crests over *Ererê* hill in the late morning.

Fig. 6.24 Rock Tower east-facing side (right taken 10:30 AM, Nov 8, 2008). A close-up of the top (top left taken 8:04 AM, Oct 30, 2009), there is art on its left side where the sun prevents green lichen from growing as photographed from atop *Ererê* depicting the northeast side of the tower (mid left taken 1:36 PM, Jan 29, 2010). About 4 red pictographs (lower left taken 8:07 AM, Jan 25, 2010): a zoomorph resembling a manatee with arms arcing downward, a concentric dot, a half arc left of the circle, and a very faded concentric dot beneath and to the left of the arc

Fig. 6.25 Sunlight reflecting high off the south side of the rock tower in late morning (top). Dot on concentric dot photographed with scale (left) at 1:36 PM, Jan 29, 2010, lower on the southeast side of tower, showing diameter is ~16 cm. The sun shadow line approaches the circle (center) at ~10:33 AM, Nov 8, 2008, and intersects the center of the dot (right) at ~10:45 AM, Nov 8, 2008

As the sun continues moving across the sky, its rays bathe the southern side of the tower, then around to the southwest, and then the west. Pictographs continue down the south side of the tower to the base—including another red dot touching a concentric dot (Fig. 6.25). A sun shadow intersects this pictograph, as the sun moves across the sky. I captured that moment by chance at 10:45 AM, but it probably occurs on late mornings most days of the year. However, the shadow line possibly intersects the middle of both dots only on one, or a few, specific (currently undiscovered) days of the year.

At the base of the rock tower is where a panel of pictographs are angled from southwest to west, while the natural ground slopes downward (Fig. 6.26). The most prominent image is another zoomorph resembling a manatee. It has a hexagonal head, diamond-shaped torso and tail, and bent-upright forelimbs. A red dot also covers part of its lower torso and tail. Two more red dots are left of it, all of which receive direct sunlight around noon. Continuing to the left is a disembodied face with circle eyes, the logogram motif, a larger red dot, and then a red dot with a bent hook at the top and a short line extending to the left, perhaps indicating the direction the sunrays continue through the day.

Fig. 6.26 Rock tower panel depicting a zoomorph resembling a manatee with arms up. The red dot on it, and dots continuing further left along the panel to the west might be tracking direct sunshine or intersecting shadow lines throughout the day, possibly on days on or near the northern solstice. Photo taken at 1:35 PM on Nov 8, 2008

The last clearly visible image in the panel is an abstract motif, but image enhancement reveals this portion of the panel has more circles. Beneath the hooked dot image, the Dstretch enhancements (Fig. 6.27) reveal a concentric circle near the center of the photo. A small circle appears right of it, and several smaller circles appear to the left of it (possibly 3 circles and a dot with a line). The last circle to the left is on a white coating made more visible on the top left side due to the enhanced photo, as well as two circular indentations—which could be natural and/or altered. Through the day, sunrays move from right to left of the photo.

At the western base of the tower, located higher up from the ground are also three altered orifices, one which has a concentric circle inside, where the setting sun illuminates the inside on days near the northern solstice (Fig. 6.28). Unfortunately, the orifices are currently too high from the ground to get precise compass readings that pinpoint the exact days and times this occurs. However, a concentric dot is positioned outside and to the right of the orifice, while a spoked half-concentric dot is positioned far beneath and to the left of the orifice. The spoked circle has a single spoke (line) angled upwards. A straight line can be drawn from that spoke through the center of the circle inside the orifice and the circle outside and to the right of it. The three circles connect in a single straight line, offering strong evidence that the artist(s) were tracking the movement of the sun through painting its sunrays as they strike the tower throughout the day.

It is perhaps this reason why Charles Hartt and Alfred Wallace believed the tower indicated some form of reverence for the sun. Direct compass measurements were not obtained due to the difficulty in reaching many of these paintings on the tower (and to not cause further erosion of the tower by climbing on it), but sunshine and shadows are clearly visible in the photographs. Some of the shadows are produced by the

Fig. 6.27 Rock tower base with Dstretch_yrd to enhance red color reveals numerous red circles on this portion of the panel beneath the abstract motif. The manatee is in the right corner and the abstract motif is at the center top edge of the photo

Fig. 6.28 Rock tower northwest to southeast side (bottom). The top sequence of photos from right to left show a concentric dot outside and one inside an altered half-circle orifice. The bottom image shows three concentric drawings, the one on the base has a straight blue line intersecting up through all three concentric drawings. All photos taken November 8, 2008, between 1:39 PM–1:42 PM local time

contours of the tower itself, which leaves future questions for investigation—whether parts of the tower were deliberately sculpted for precision. There are no known images on the northeastern side of the tower, which mostly remains in shadow.

Returning to the logogram in the panel on the southwest base of the rock tower, another motif was revealed with the photo enhancement. Positioned right below the logogram is the cornu spiral motif. These two motifs, along with the abstract design, which is further to the left, appear to be part of a linear sequence. There is a sense that some instruction or explanation is encoded in these symbols, particularly as it relates to the rock tower and probably the sun.

A cornu spiral motif also appears low on the wall beneath the sun pedestal pictograph at the beginning of the Serra da Lua trail, and there is a large cornu spiral beyond the rock tower on the continued path toward Serra do Sol. It also appears at a few other rock art locations throughout Monte Alegre (Fig. 6.29). The cornu spiral is perhaps an indicator of "beginnings." Perhaps it signifies the beginning of a rock art sequence, or perhaps it signifies a solstice sequence. We should not rule out that it signifies more than one thing, as well.

Fig. 6.29 Cornu spiral motifs at the rock tower panel (top), the base of the wall at the sun pedestal panel (bottom left), and at the continued rock art sequence heading in the direction of *Serra do Sol* (bottom right)

References

Capitaine, N., P. T. Wallace, and J. Chapront. 2003. Expressions for IAU 2000 precession quantities. *Astronomy & Astrophysics* 412 (2): 567–586.

Collins, G. W., II, W. P. Claspy, and J. C. Martin. 1999. A reinterpretation of historical references to the supernova of A.D. 1054. *Publications of the Astronomical Society of the Pacific* 3 (761): 871–880.

d'Abbeville, P. C. 1614. *Histoire De La Mission Des Pères Capucins En L'isle De Marignan Et Terres Circonvoisines Où Est Traicté Des Singularitez Admirables & Des Meurs Merueilleuses Des Indiens Habitans De Ce Pais Auec Les Missiues Et Aduis Qui Ont Este Enuoyez De Nouueau*. Paris: de l'imprimerie de François Huby ruë St. Iacques à la Bible d'or et en sa boutique au Palais en la gallérie des Prisonniers.

Davis, C. 2023. Amazonian animism: Natural world annotations in paleoindian cave art. In *Amazon ecosystem-past discoveries and future prospects*. IntechOpen.

Fabian, S. M. 2001. *Patterns in the sky: An introduction to ethnoastronomy*. Prospect Heights: Waveland Press, Inc.

Graham, D. W., and E. Hintz. 2010. An ancient Greek sighting of Halley's comet? *Journal of Cosmology* 9:2130–2136.

Kelley, D., and E. Milone. 2011. *Exploring ancient skies: A survey of ancient and cultural astronomy*. 2nd ed. New York: Springer.

Needham, J. 1981. In *The shorter science & civilisation in China: 2: An abridgement of Joseph Needham's original text*, ed. C. A. Ronan. Cambridge University Press.

Nimuendajú, C. 1946. *The Eastern Timbira*. Berkeley: University of California Press.

Pankenier, D. W. 2013. *Astrology and cosmology in early China: Conforming earth to heaven*. Cambridge: Cambridge University Press.

Reichel-Dolmatoff, G. 1982. Astronomical models of social behavior among some Indians of Colombia. In *Archaeoastronomy in the new world: American primitive astronomy*, ed. A. F. Aveni and G. Urton, 165–181. Cambridge University Press United Kingdom.

SunEarthTools.com. 2024. Annual Sun Path. 2024 ed.

Chapter 7
Stone Tools and Artifacts at Ererê

Serra da Lua Surface Artifacts

Because *Serra da Lua* is a pictographic trail halfway up the cliffs of *Ererê* hill, there is little soil accretion to excavate, and no clear spot along the 240 meters that stands out as a suitable location for excavation. There are a few "corners" formed by two adjacent walls here and there along the trail, including a narrow cave in one "corner," but the cave is too narrow for people to have entered—or for us to excavate. It is more like a crevasse between two cliff slabs. Thus, no apparent location along the trail was suitable for excavation.

However, the frequent visits to the paintings gave my helpers and me plenty of opportunities to notice artifacts on the surface. We often ventured beyond the trodden paths and stayed on the hill far longer than most visitors, including locals, who rarely go into the hills except as guides for tourists. We were fortunate to find three lithics lying on the surface, partially concealed by underbrush and lodged next to boulders. Surface finds do not provide an opportunity to date the artifacts, but their attributes and characteristics can (and will) be compared to dated artifacts in later chapters.

The three surface artifacts were collected at *Serra da Lua* near the beginning of the pictographic trail (Fig. 7.1), and all were lithic tools with evidence of human use. The first lithic tool (Fig. 7.2) is a gray, maroon, and purple banded blade (a lithic "blade" is a thin flaked rock that is at least twice as long as it is wide). Its banded colors are comparable to the material composition of part of the cliff wall, so it is strongly suggestive that it was quarried off this portion of the wall during the Stone Age period. However, the blade appears to have been used to strike or rub against a red-colored object sometime after it was detached from the wall.

Although this is a surface find, it suggests that ancient people may have been drawn to *Monte Alegre* as a potential lithic quarry. It also raises the question of whether they sought and used naturally colored rocks as tools specifically suited for

C. S. Davis, *Before Sunset*, Conflict, Environment, and Social Complexity,
https://doi.org/10.1007/978-3-031-93373-8_7

Fig. 7.1 Beginning of Serra da Lua's pictographic trail at the southern start. Moving from south toward the north would move from the right side of the image to the left. The purplish-maroon-colored rock on the right side is in the background, the southern solstice image is in the center (shown face-forward later) and a protrusion of the wall that was painted with red circular "eyes," curvilinear eyebrows, nose, and mouth are visible in the foreground on the left side of the image. Another human-like face can be discerned from the void spaces in the rock at the very left edge, but this was not noticed at the site and could just be a pareidolia effect in this photo. The red painted face, however, is demonstrably intentional

preparing pigments. This is a clue that animism, possibly akin to imitative magic, may have been applied. A more practical explanation, however, is that the paint was prepared directly at the drawing site—halfway up the hill along the pictographic trail—instead of at a base camp or cave.

The second surface artifact collected at *Serra da Lua* (Fig. 7.3) has a cleaver-like distal edge, a proximal tip that is darkened and stained red, and an overall surface that appears artificially white-painted. A reddish-orange painted line is visible on the flat dorsal surface near its proximal edge. The darkened edges are partly due to dirt, but there is also evidence of dark soot on the proximal and distal surfaces, suggesting the tool may have been used against charred materials.

The third surface artifact collected at *Serra da Lua* was labeled as a *cortex flake* (Fig. 7.4) because the dorsal side displays a very dark gray surface, possibly due to being the outer, weathered cortex of the parent rock it was flaked from or due to later varnish and wildfire soot. In contrast, the ventral side reveals a bright white polished surface with red and yellow marks. For this flake, the red and yellow pigments appear to be embedded in the rock due to the high degree of polish. However, both pigments were likely applied as paint and then heavily polished. While some crystals may have been naturally red, the pigment seems to have been expertly applied.

Fig. 7.2 *Serra da Lua* purple/maroon surface artifact. The dorsal side (**a**), lateral edge (**b**), and ventral side (**c**) exhibit the chisel-like shape. The lithic was discovered on the surface along with another lithic (**d**) near the southern solstice image. Its gray, maroon, and lavender banded appearance matches the material composition of the cliff wall near here. However, the red mark visible at 30× magnification (**e**) appears to be from contact with red pigment after it was knapped

Excavation Above Serra da Lua

Near the rock tower, a different pathway can be taken to reach the top of the hill. It is not a formal trail, but it is the easiest way to climb past the cliff walls to reach the top of the hill on the western region of *Ererê.* We hiked atop the cliffs, above *Serra da Lua*, and dug some shovel test pits at a location with relatively flat ground (Fig. 7.5) approximately 181 meters above sea level (64 meters higher than the start of the Serra da Lua pictographic trail). The test unearthed rocks that looked potentially altered by humans, including one with unnaturally red stains.

The location atop the hill may have attracted Stone Age people because once ascended, it is easier to move atop the hill to the *Serra do Sol* location than to move through dense overgrowth along narrow and angled ridges halfway up. It also

Fig. 7.3 *Serra da Lua* white surface artifact. The dorsal and lateral side of this bulbous flake has a cleaver-like distal tip (left). At 30× magnification (right), a linear red feature near its proximal edge is caused by a dull orange/red mastic between rock crystals. The rest of the lithic not crusted with blackened dirt is a bright white, polished surface. This white coating too appears to have been applied

provides good rest stop just after ascending the hill, and the higher elevation flat ground still provides a view of *Serra do Sol* in the distance (Fig. 7.6). Since the concentric dot faces the northern solstice, having a line of sight to it is like having a compass marking atop the hill. The location is also protected on most sides by rocky relief (revisit Fig. 7.5). If ancient people traveled through this region, it was possible that they could have camped out on this flat, partially sheltered ground—protected on three sides from wind and other dangers—while in view of the rock art.

Alternatively, the location itself could have been used as a "kill box" long ago, where game animals could have been chased into the narrowing terrain, so two side-by-side 1 × 1-meter units were excavated here to explore those theories (Fig. 7.7).

The excavations revealed no animal bones, only rocks found beneath tree and shrub roots. No ceramics, no charcoal from campfires, no apparent animal or plant

Fig. 7.4 *Serra da Lua* surface cortex flake artifact. The (**a**) lateral edge and (**b**) dorsal side show a very dark gray surface due to either being the outer, weathered cortex of the parent rock it was flaked from, or due to later varnish and wildfire soot. The (**c**) ventral side reveals a bright white surface that contains a red and yellow region with a high degree of polish visible at (**d**) ×20 magnification

processing, and no apparent trash middens—there was no evidence of a camp. There was also relatively thin soil accretion (92 cm depth to bedrock), but quite a few interesting lithic fragments were unearthed, particularly starting at about level 6 (56 cm depth) through to level 9 (86 cm).

Fig. 7.5 Above *Serra da Lua* shovel pit testing of relatively flat ground sheltered by terrain revealed potential lithic artifacts (inset), the center rock has reddish stain or paint. Brazilian architect and research assistant, Carmen Silvia, pictured in photo

Fig. 7.6 Above *Serra da Lua*, the test pit location was within sight of *Serra do Sol* to the east (left). A zoomed-in photo of the >1-meter-large concentric dot painting facing the northern summer solstice (right)

Fig. 7.7 The excavation on top of *Ererê* consisted of two side-by-side 1 × 1-meter-square units (top). The profile view (bottom left) of unit 1's west wall, and the plan view of level 9 (bottom right) for unit 1 and part of unit 2 visible at the top of the photo. In both photos, the yellow leveler points north–south. Author pictured in photo

Some of the lithics have red lines, apparently drawn intentionally by humans (Figs. 7.8 and 7.9). Some rocks appear to have been utilized, possibly to apply or remove paint off other surfaces (Fig. 7.10). Some lithic knapping was evident, but it could not be ruled out that some flakes fractured off due to wildfires (Fig. 7.11). Alternatively, they could have been intentionally fire-cracked rock.

Fig. 7.8 Lithic artifact excavated from unit 1 level 8 "Above" *Serra da Lua*. The lithic fragment (left) exhibits a narrow red line along its transverse edge, clearly visible under 30× magnification (right)

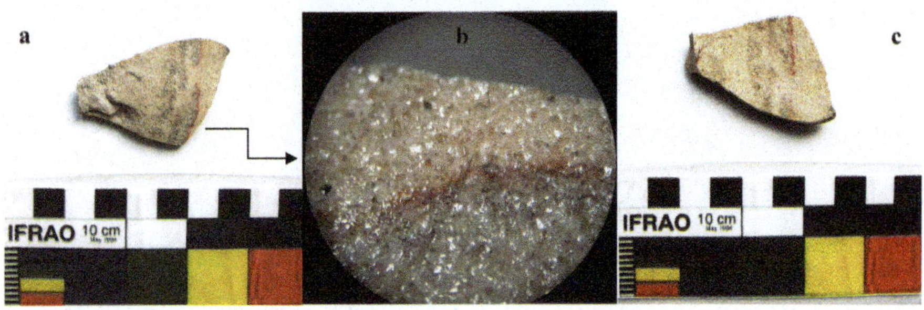

Fig. 7.9 Lithic artifact excavated from unit 2 level 8 "Above" *Serra da Lua*. The lithic flake exhibits a narrow red line painted along its dorsal side flake scar (**a**). The red pigment is noticeably narrow and precise under 30× magnification (**b**), which also demonstrates its polished surface to the point that the red color appears to be embedded in the rock. The painted red line continues halfway-down the ventral side (**c**)

Fig. 7.10 Artificially pigmented lithics excavated from unit 2 level 6 "Above" *Serra da Lua*. The lithic in top left image was observed under 30× magnification showing (below left) red pigment and (below right) orange/yellow pigment on different sides of the same tip

Some artifacts also exhibit highly polished surfaces (Fig. 7.12), possibly from rubbing activity with softer substances, such as skin or animal hide. There remains some uncertainty about whether the artifact from Unit 2, Level 8, was deliberately knapped or if it flaked off due to spalling (induced by fire). The concave dorsal face, visible from the distal viewpoint (d), seems unlikely to have formed spontaneously by fire and appears more symmetrical than what is typically produced by the Hertzian cone force of direct hammer percussive knapping. It is possible that the lithic piece was abraded into this concave shape. The highly polished surface, visible under magnification (b), strongly suggests intense human modification regardless.

Fig. 7.11 Lithic fragments excavated from unit 2 level 7 "Above" *Serra da Lua*. Many of these fragments could be natural breakages or fire-cracked (spalled) flakes from ancient wildfires. However, some have polished surfaces (all the rocks in the second row), and some have pigment (second from left on top row, and sixth from left on bottom row)

The lithic artifacts in level 8 of unit 2 yielded the clearest evidence of human activity (Fig. 7.13) often with an artisanal focus. The middle west region of the excavation pit contained a lithic cluster left behind by the knapping of one or a few flaked cores. Some of the flakes in the cluster (the bottom row) have residual red pigment.

A burin-like flake (Fig. 7.14) from the lithic cluster was excavated from the middle west region of unit 2 level 8, and it has a red-stained edge and a beige concretion on its tip. A similar type of concretion was discovered on lithics from the lowest levels of the *Painel do Pilão* excavation in the next chapter as well. During microscopic analysis, a single hair was also noticed stuck to the concretion at the tip, but the hair was not noticed during excavation. I could not be certain whether the hair was ancient, or if it occurred from brushing during excavation or cleaning the artifacts in the laboratory afterwards.

A blade-like artifact (Fig. 7.15) from the lithic cluster was also excavated from the middle west region of unit 2 level 8. Its dorsal and ventral sides both have reddish-yellow staining covering their entire upper (proximal) half. The blade itself appears to have broken at both ends, but the staining does not seem likely to have occurred by use against a reddish substance. This staining pattern is more likely to have occurred by intended coating of the lithic with red pigment.

Fig. 7.12 Polished lithic artifact from unit 2 level 8 "Above" *Serra da Lua*. The dorsal side (**a**) at 30× magnification of the central region shows a highly polished surface (**b**). The ventral side (**c**) shows a smooth concave rectangle, and the distal edge allows viewing of the smaller concave scoop of the dorsal mid-region (**d**) and a trapezoidal shape overall

The longest knapped flake was a ~12-cm-long blade (Fig. 7.16) presumably wrought from a single percussive strike of what must have been an impressive-sized core. Yet the ventral side of the flake is relatively thin, flat, and smooth. There is little evidence whether it was used or not, but the intention may have been as a wood scraper to smooth or shave wooden posts.

Another recognizable artifact from unit 2 level 8 are the two ochre nodules (Fig. 7.17). This ochre is possibly hematite-rich clay, or goethite, and they must have been transported to the top of the hill, presumably for intentional paint activity. The discovery of them in just about the lowest stratigraphic level, coinciding with the lithics found with paint on them, combines to provide the strongest evidence that painting activity occurred in the earliest period of human habitation at *Ererê*.

Fig. 7.13 Lithic flake cluster excavated from the middle west region of unit 2 level 8 "Above" *Serra da Lua*. The cluster of flakes is displayed on their dorsal (left) and ventral (right) sides. The first three flakes from left in the bottom row have evidence of pigment stains

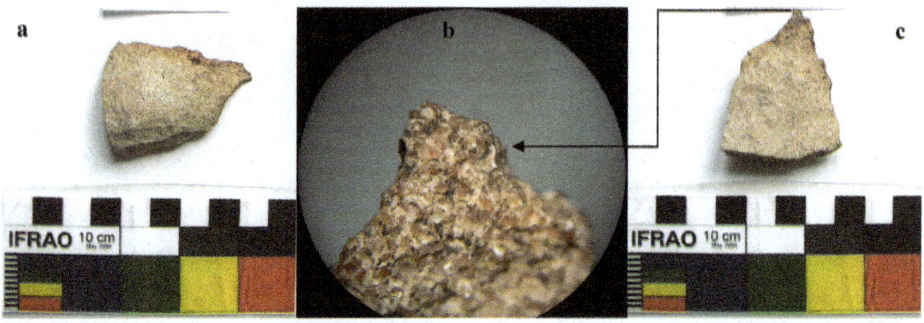

Fig. 7.14 Burin-like artifact is the first from left in Fig. 7.13 lithic cluster. Its dorsal side has some red staining (**a**). The dorsal tip at 30× magnification shows a hardened spur with a beige concretion and possibly a hair or fiber lodged in it (**b**). The ventral side showing the area that was magnified (**c**)

Fig. 7.15 Blade-like artifact is the third from left in Fig. 7.13 flake cluster. Its dorsal side (left) has reddish-yellow staining on its entire upper (proximal) portion. Its ventral side (right) is similarly stained on the upper proximal portion. A curved scar holds more concentrated red pigment

Fig. 7.16 Lithic blade excavated from unit 2 level 8 "Above" *Serra da Lua*. The blade is long and thin, with a lengthwise flake scar on its (**a**) dorsal side. Part of the (**b**) ventral side can be seen from its lateral edge, and dorsal side is visible from this (**c**) lateral edge angle

Fig. 7.17 Ochre pigment nodules excavated from unit 2 level 8 "Above" *Serra da Lua*. The two mostly red nodules (left) with trace yellow pigment under ×30 magnification (right)

Level 9 is the lowest excavation level of both units, both reaching bedrock. Those levels too yielded lithics with red marks on them. One rock is mostly covered in red pigment, one has bright red marks, and a third flake is stained on its tip (Fig. 7.18). Through relative dating, these flakes are likely the oldest artifacts at this location. Although no sufficient carbon was excavated to perform radiocarbon dating of these layers, a comparison of these artifact attributes with those discussed in the following *Painel do Pilão* chapter suggests contemporaneity. Lithics and pigments were also found at the lowest levels there, indicating a potential overlap in time.

Serra do Sol Lithic Cache

In addition to the artifacts excavated from above *Serra da Lua*, a cache of lithic artifacts was discovered in an orifice (Fig. 7.19) just a few meters from the Northern Solstice panel at *Serra do Sol*. A faint red paint streak marks the mouth of the hole, indicating human alteration. Inside, 23 artifacts were pulled from the fist-sized hole, but concern of encountering a snake nest prevented a complete exploration of the hole or anything else within.

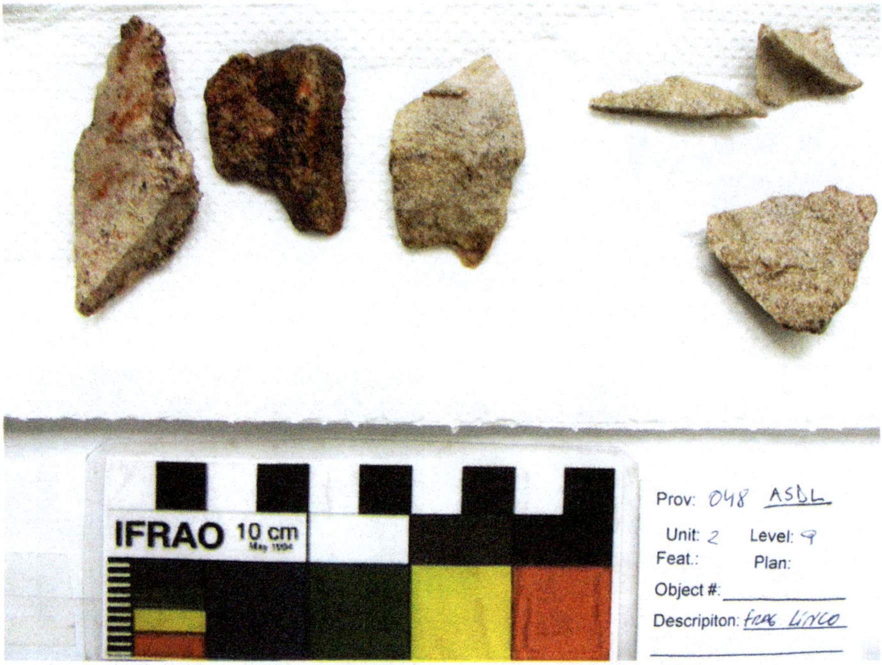

Fig. 7.18 Lithic fragments from unit 2 level 9 "Above" *Serra da Lua*. The three flakes on the left all have red pigment on their surfaces or tips

Many of the lithics have permanent duo-textured surfaces—one part being smooth and light, another part being rough and dark—as if they were partly exposed to a vastly different heating or weathering process (Fig. 7.20). The strongest theory is that the dark, rough texture is where the stone tools were hafted and lashed onto a wooden shaft, but the wood and binding material have long since decayed. This theory does not, however, satisfactorily explain why the possibly hafted portion of the rock would appear to have been chemically altered (Fig. 7.21). Perhaps some form of pitch using resinous wood was made as a "glue" to bind the rock to wood, but no evidence of any sort of mastic has been recognized with microscopic analysis. Perhaps a precipitate deposited on the lithics in the hole from burned rock and iron or manganese-rich soil surrounding them. Some artifacts have black micro spherules which could indicate this kind of process (Fig. 7.22). However, more flakes have distinct boundaries where just the dark side seems to be "scorched," with part of the normally smooth rock matrix missing in some cases, leaving behind granular—or even acicular—crystalline quartz sandstone (Fig. 7.23e).

Some of the lithic artifacts were analyzed with pXRF to determine chemical elements of some of their surface features. Sample #2 was the most distinct because it had a stark white coating on its dorsal side, and only minimal "scorching" on its ventral side. This artifact is evidence suggesting that the "scorched" condition of many of the other flakes is unlikely to have occurred when all the artifacts were in

Fig. 7.19 Lithic cache at *Serra do Sol*. Twenty-three lithic artifacts were discovered (top right) at the Northern Solstice panel (left) in a fist-sized hole with a painted red line above it (lower right). Many of these lithic flakes had permanent duo-textured surfaces. Author pictured in photo

the cache. We would otherwise expect to find similar scorching on this sample. Instead, the XRF revealed elevated levels of phosphorus (70,000 ppm or 7%) found on the white-coated surface (Fig. 7.24).

The beige-colored tip of the dorsal side, which lacks the coating, was also analyzed, and found to contain only 1.4% phosphorus, so the white coating is unlikely to have formed naturally from salt precipitates (Fig. 7.25). It appears to have been purposeful. From the XRF analysis, we can conclude that the red paint mark on its

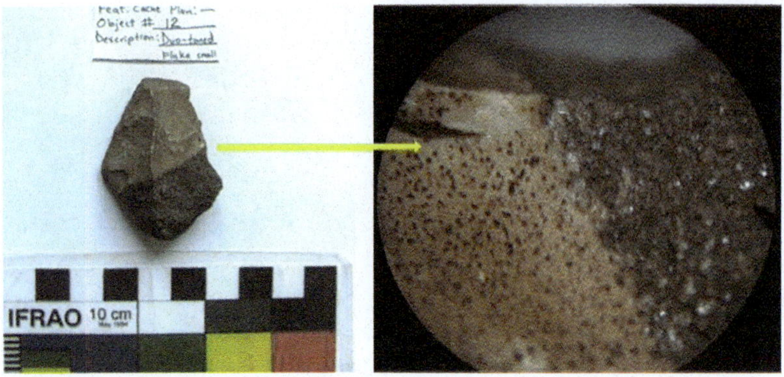

Fig. 7.20 Lithic artifact #12 from cache at *Serra do Sol* "Northern Solstice" panel. The small flake has a "scorched" appearance and texture at a definitive line along its mid-region on the dorsal (left) and ventral side. Under ×20 magnification (right), the dividing line preserves the different crystalline texture and darker color

Fig. 7.21 Lithic artifact #3 from cache at *Serra do Sol* "Northern Solstice" panel. The convex flake has a reddish swath on its dorsal spine and a blackened appearance along its lateral edge (top left), made more distinct where a white residue borders the dividing line under ×20 magnification (bottom left). Its ventral side is also blackened and lacks a reddish hue (top right) except under ×20 magnification (bottom right), which reveals light red (paint?) in discontinuous areas underneath patchy, dark micro spherules and flecks

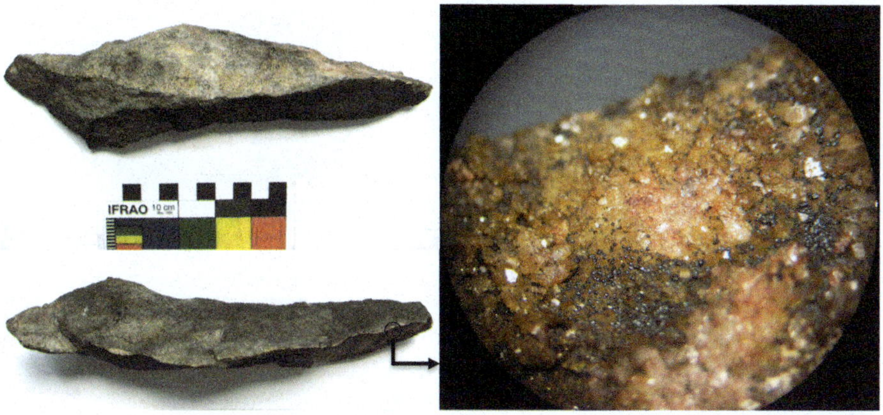

Fig. 7.22 Lithic artifact #1 from cache at *Serra do Sol* "Northern Solstice" panel. The large flake has a "scorched" appearance and texture at a definitive line along its transverse axis mid-region on the dorsal side (left top). Along most of the ventral side (left bottom), it appears blackened. Magnification at ×30 (right) reveals some of the blackened appearance to be caused by black micro spherules, presumably of burned, manganese or iron-rich staining, but unproven

distal dorsal ridge is iron rich, and the white coating is rich in phosphorus. Both colors were intentionally applied to the lithic.

It is not entirely clear, however, that white coloration is caused solely by phosphorus, since sample #4 showed no phosphorus on its light-colored dorsal ridge (Fig. 7.26), but it did contain phosphorus everywhere else. The sample also had a fiber embedded in its rough region (Fig. 7.27). Although the fiber is unknown, it is rigid (not a root, but perhaps petrified?). Its presence suggests the lithic was not subject to open flame since deposited.

Red marks on other samples besides sample #2 also register higher iron. Sample #6 has the most vibrant red marks, but its increase of iron does not proportionately reflect that (Fig. 7.28). The red color must have been mixed with something organic. However, its "scorched" region coincides with a decrease of ~1000 ppm sulfur and titanium, while its white coating returns a high value of phosphorus (Fig. 7.29). The XRF analysis of sample #10 further reinforces the finding of a phosphorus-rich light surface and the reduction of phosphorus and sulfur in the "scorched" region (Fig. 7.30). Perhaps a "coating" helped preserve the rock, acting to prevent normal erosion of coarse sandstone rock.

Fig. 7.23 Lithic artifact #9 from cache at *Serra do Sol* "Northern Solstice" panel. The dorsal side of the dark brown flake (**a**) has a mid-dorsal flake scar (**b**) defined by a blackened curvilinear border seen under ×20 magnification (**c**). Its ventral side (**d**) has a "scorched" proximal lateral edge distinctly darker and more jagged than the smoother ventral surface under ×20 magnification (**e**). Small black specks throughout both regions, which may have been caused by melted soil, indicates the edge was caused by a different action

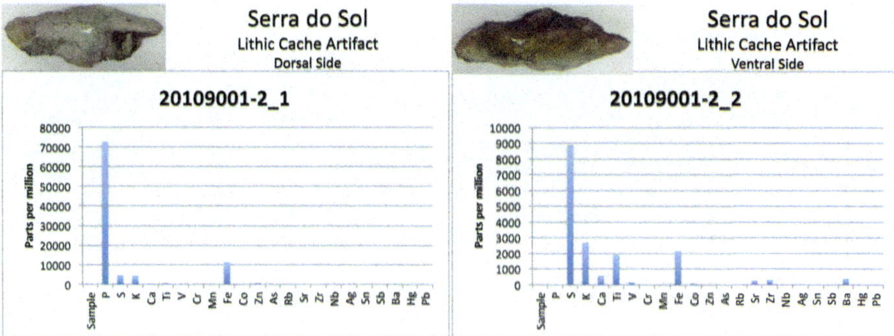

Fig. 7.24 *Serra do Sol* lithic cache sample #2. XRF analyses were performed at spot indicated by the tip of the foil arrow. Dorsal side (left) shows ~7% phosphorus and ~1% iron content. The ventral side (right) shows ~0.9% sulfur and 0% phosphorus

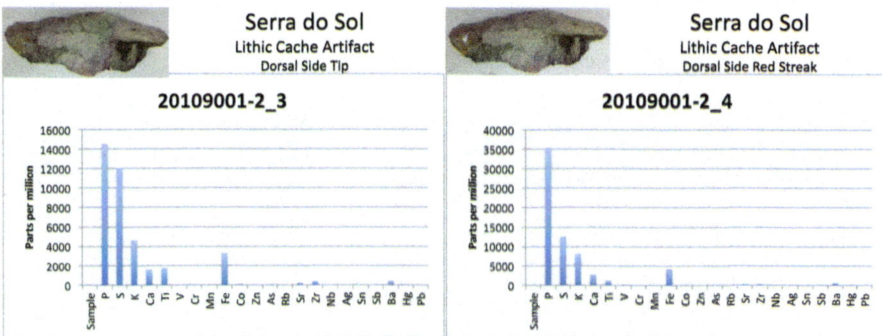

Fig. 7.25 *Serra do Sol* lithic cache sample #2. XRF analyses were performed at spot indicated by the tip of the foil arrow. Dorsal tip, plain surface (left) shows ~1.4% phosphorus and 1.2% sulfur. Dorsal ridge, red streak (right) shows higher phosphorus (3.5%) and iron (~0.5%)

Fig. 7.26 *Serra do Sol* lithic cache sample #4. XRF analyses were performed at spot indicated by the tip of the foil arrow. Smooth ventral tip (left) shows ~1.3% phosphorus content. Dorsal side (right) shows no phosphorus content. Both sides have ~0.3% sulfur

Fig. 7.27 *Serra do Sol* lithic cache sample #4. XRF analysis was performed at spot indicated by the tip of the foil arrow (left). Rough ventral side shows ~1.5% phosphorus and 0.8% sulfur content (graph). The dorsal side (top right) has a darkened mid-section seen at ×30 magnification which revealed a single strand "fiber" embedded into the crystal (bottom right). The fiber is unknown but presumably ancient because it was embedded in the rock

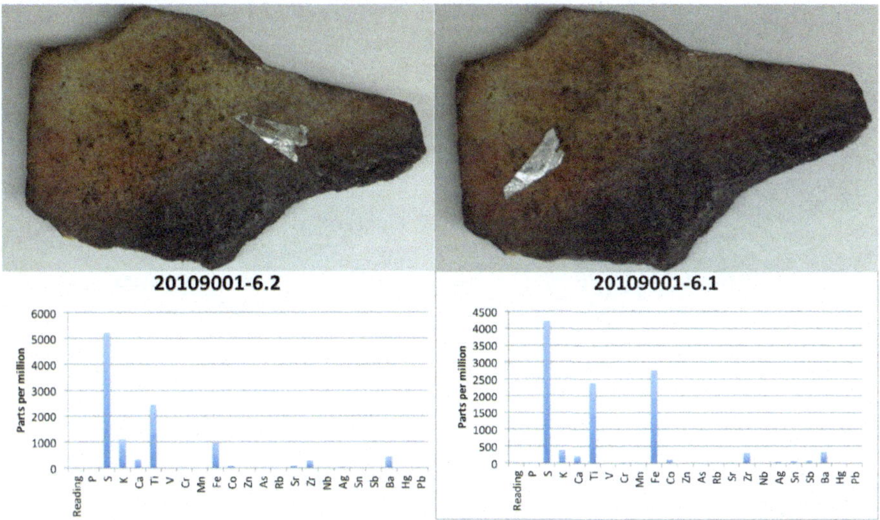

Fig. 7.28 *Serra do Sol* cache sample #6. XRF analyses performed at spots indicated by arrow tip. Dorsal side unpainted surface (left) shows less iron content (0.1%) than the surface with the red streak (right), which shows an increase of iron by ~1700 ppm

Fig. 7.29 *Serra do Sol* lithic cache sample #6. XRF analyses were performed at spot indicated by the tip of the foil arrow. Dorsal side "scorched" surface (left) shows less titanium content (0.1%) than in the previous two surface locations. Ventral side shows high phosphorus content (1.4%) compared to the dorsal side

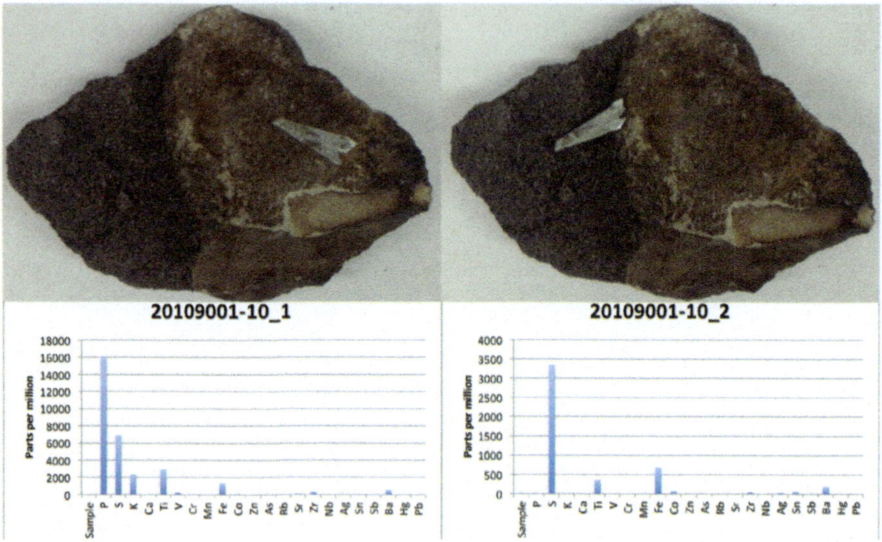

Fig. 7.30 *Serra do Sol* lithic cache sample #10. XRF analyses were performed at spot indicated by the tip of the foil arrow. Dorsal side plain surface (left) shows ~1.6% phosphorus and ~0.7% sulfur. Dorsal side burned surface (right) shows no phosphorus or potassium, and less sulfur than the non-burned half

Chapter 8
Excavation and Investigation at Painel do Pilao

Introduction

Painel do Pilão is an open-air pictographic rock art shelter with an elevation of about 135 meters above sea level on *Serra da Paituna* hill in the *Monte Alegre* rainforest, which borders the Amazon River. Its research and excavation data present a compelling sequence of both site and tool artistry, and perhaps secondary use. A semi-subterranean shelter there may have been initially used as a sandstone quarry, but one wall seems to have also been terraformed. Theories about this wall are explored in this chapter and the next. Although a new theory is introduced in the next chapter, the theory explored in this chapter is whether the lithics retrieved from excavating the shelter space, and the terraformed wall unearthed within it, indicate quarrying activity alone, or instead stem from the formation of a planned stage or platform from which to view the rock art painted above the shelter.

Over the ages, rocky sediment continued accumulating from erosion and the few subsequent times of human occupation, preserving a progression of lithic artifacts. The oldest lithics were perhaps originally selected due to their naturally colored inclusions, but quickly thereafter, the styles change to an apparently intentional coating and drawing in paint of red, yellow, and whitish hues. The combined contexts of the lithics, the rock art, and the reshaped wall speak to significant levels of artistic and pre-scientific efforts hitherto unaddressed by research on Paleoindians during the peopling of the Americas.

The sandstone schist outcrop is ~14 meters high total, with the mostly red and yellow ochre pictographs spanning approximately 3.35 meters high by 4.81 meters long. They are clustered on the lower two-thirds of the outcrop, painted on surfaces facing outward, downward, or even upward, as the painted wall curves around to form a virtual ledge that divides the art among two levels (Fig. 8.1). Art on the lower level is clearly visible and accessible from a large horizontal platform stage that forms the base of the outcrop (and is also the east sidewall of the semi-subterranean

C. S. Davis, *Before Sunset*, Conflict, Environment, and Social Complexity,
https://doi.org/10.1007/978-3-031-93373-8_8

Fig. 8.1 *Painel do Pilão* painted outcrop showing the lower- and upper-level rock art, the stage, and the lower excavated shelter enclosure

excavated shelter discussed next). Art on the upper level could have only been painted by ascending the ~2-meter-high ledge, which is the only position from which art on the "floor" of that ledge can be viewed. However, art on the wall and ceiling of the upper level can still be seen from the stage.

The platform/stage itself is part of a massive boulder jutting out from beneath the outcrop just described. The massive boulder has an ovoid shape and is upraised relative to the ground around it, with the rock shelter located on its northwest edge (Fig. 8.2), and the vertical painted outcrop forms its roof. The shelter is only accessible by walking around the north-facing edge of the massive boulder, where the ground slopes upward almost like a ramp (Fig. 8.3). Only the edge of the massive boulder located within the shelter (under the painted ledges) had been reshaped into a flat, uniform, vertical wall; and the lower one/third of this wall was only recently unearthed during the excavation.

Human efforts to transform landscapes are often associated with sedentary communities, with only recent hunter-gatherer exceptions coming to light. *Göbekli Tepe* (Caletti 2020) and *Nevali Çori* (Tobolczyk 2016) are notable exceptions in the Old World, while the *Tampa Bay* region (Randall and Sassaman 2017) and the *Salish* area on the Pacific Northwest coast (Grier et al. 2017) are possible examples in the New World. Similar to those cultures, evidence for Paleoindian habitation at *Monte*

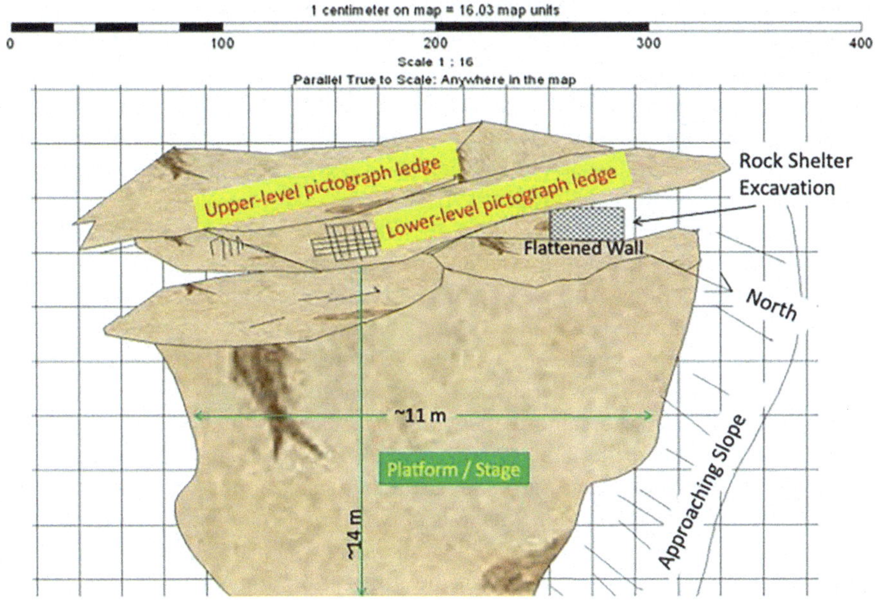

Fig. 8.2 Site map of *Painel do Pilão* looking top down on the excavation, shown in crosshatch in the upper right corner. Magnetic north marks the excavation datum point. The "grid" painting and line and circle painting are represented on the vertical rock surface. The long "tongue" is the platform/stage, ~14 by ~11 meters

Fig. 8.3 Painel do Pilão (Left photo) view of slope and platform stage to the left side, while facing the rock shelter entrance, which is an immediate left turn at the top of the terrain. (right photo) View of ground looking down slope (northeast) from the rock shelter, in the distance is a ~30-cm-tall black bucket holding plastic, black tarp. To the right side, in the foreground, is the north corner of the platform (stage), just outside the rock shelter

Alegre has consistently been interpreted as seasonal (Roosevelt et al. 1996). By no means is there evidence of large-scale earthwork transformations at *Monte Alegre* anywhere near the scale of *Göbekli Tepe* and the other sites. Still, the possibility of small-scale rock carving for *archaeoastronomy* utility instead of religious and/or artistic symbolism is notable and similar to the other sites mentioned above. More associations of this kind of activity at the end of the Ice Age are increasingly being discovered (De Lorenzis and Orofino 2014).

Some of the rock art produced by the ancient people hint at notational recordings of the sun along the horizon throughout the year (Davis et al. 2017). So, this site supports new perceptions of Paleoindians and of hunter-gatherers worldwide, suggesting that some groups began transforming habitats before permanently settling—and that their motives correlate with pre-scientific attempts to record natural reoccurrences that predict or herald seasonal phenomena.

The common perception among Paleoindian research is that material culture was mostly produced for utilitarian efforts related to subsistence and habitat adaptability. However, the *Painel* excavation suggests that sensory and conceptual motivations were a stronger factor than what has previously been considered. Some of the recovered lithics show signs of use while others show signs of artistry, and yet, the lasting impression that remains is that the greatest amount of labor went toward landscape modification of the northeast shelter wall—the one wall that composes part of the platform stage.

In a previous publication (Davis et al. 2017), the nature of the rock art at *Serra da Lua* was argued to have solar themes aligned to horizon sightings from northern to southern solstice sunsets—a solar year. The rock art at *Painel do Pilão* is believed to also track the sun, especially at solstice and near equinox intervals, but seemingly as it relates more so to sunrises. This archaeoastronomy research is presented in the next chapter. This chapter focuses on the potential tool use and landscape modifications carried out by Paleoindians at *Painel do Pilão*, based on an archaeological excavation of the semi-subterranean rock shelter (Fig. 8.4), and the stone tools that were unearthed there.

Excavation

During the excavation of *Painel do Pilão*, 1688 lithic artifacts were removed. The excavation was carried out on the (approximately 165 cm × 185 cm rectangular) floor of a semi-subterranean rock shelter excavated to bedrock in 2011. The shelter is positioned at the base of a silicified sandstone rock outcrop. Here, lithic artifacts mostly composed of sandstone in various stages of slight geological metamorphism were retrieved from sandy loess soil containing human-produced gymnosperm wood charcoal, radiocarbon dated as far back as the terminal Pleistocene period. The various stages of sandstone metamorphism included silicified lithics, chalcedony, and opaque quartz. Many lithics were simple flakes, blanks, debitage, or nodules, but 110 samples displayed cursory manufacture and use, and all of them were analyzed under a microscope at 20 to 30 times magnification. Doing so revealed

Fig. 8.4 *Painel do Pilão* shelter floor, approximately 1.6 × 1.8-meter area, excavated by the author in 2011. The left (northeast) wall forms the base of a platform stage from which one can view the rock art

unanticipated decorative surface features on 43 of them, which are consistent with lithic features reported from *Serra da Lua/Serra do Sol*:

1. A cream to off-white coating or slip most consistently on the dorsal side of flakes, but at times also appearing on the ventral side of a few samples too.
2. Coated and non-coated samples with polished surfaces, most consistently on the dorsal side of flakes, but additional polished surfaces sometimes appear on the ventral side too—partially or in patches. Most polish appears to be deliberate, but some patchy sections could be the consequence of repeated functional tool use indicative of "scraper" activities.
3. Micro evidence of conglomerated or scraped-off surface residue, localized micro edge chippings, eraillure fractures, and bidirectional abrasions infer functional tool use.
4. Some flakes (mostly in the lowest levels) display a mid-dorsal flattened region (perhaps from intentional flaking or fluting), possibly to improve hafting/lashing, or generally for ease in holding the lithic without incurring damage to the hand. This theory is loosely supported by dark-line residue on unwashed lithics, suggesting decayed cordage.

Extra analysis was made to explore several null hypotheses—that tool use and residue/coating were coincidental or natural. Most of these considerations were rejectable outright, however, the fifth null hypothesis could not be fully dismissed. The null hypotheses considered were as follows:

1. Coatings were natural accretions or varnish that formed without human involvement.
2. The coatings naturally accreted from unintentional and unrecognized human involvement (for example: if paint from above could have spattered onto rocks in the shelter below).
3. Coatings were from intentional human activity (painting a wall), but the "lithics" were unintentionally created by the intentional or unintentional chipping away of that wall.
4. Coatings were from intentional human dumping or flinging of paint haphazardly into the shelter below.
5. The rocks were coated and/or painted beforehand, while on a wall or other surface, perhaps to mark those wall locations for removal.

Radiocarbon Dates

Five radiocarbon dates of gymnosperm wood charcoal taken in situ from the lowest levels (levels 6–8) of the excavation at depths of 82 cm, two at 87 cm, one at 90 cm, and one at bedrock were determined by accelerator mass spectrometry. Three additional samples were analyzed a few years later using the same methods for the higher levels (levels 2–5). All samples were excavated during the same season, but funds to analyze the additional carbon samples came later.

The highest dated charcoal sample (*seemingly gymnosperm wood*) was excavated at a depth of approximately 18 cm within *level 2*, which was otherwise devoid of lithics or any other cultural materials. It is important here to note that none of the levels, including the modern surface layer, contained ceramics. The second sample (*seemingly from a charred tuber*) was excavated from *level 4* at a depth of ~60 cm, directly beneath two lithics resembling a sickle and an end scraper; use-wear analysis of both lithics suggest those functions. The third sample (*seemingly angiosperm wood*) was excavated from *level 5* at a depth of ~72 cm. These new dates are combined with 5 previously published dates, and all are presented here on Table 8.1, offering a more complete overall excavation-layer chronology. A *gigapan panorama* of the excavation can be found at http://gigapan.com/gigapans/81478.

The dates corresponding to the higher cultural deposits give three episodic dates of *six, four, and one thousand calibrated* (and rounded off) *years BP*, with the most recent date not having context to cultural remains. The dates corresponding to the lowest cultural depositions span a total time range from *11,281 to 10,931 uncalibrated years BP* (*13,236–12,755 calibrated years BP* at 2σ using *IntCal20*), termed the *early Paleoindian Period*, and are contemporary with western *Clovis* at *13,300–12,800 cal BP* (Haynes 2002; Holliday 2000), both of which end when the *Younger Dryas cold period* of *12,800 to ~ 11,700 cal BP* (Cheng et al. 2020; Rasmussen et al. 2006; Reinig et al. 2021) begins. Of particular significance is that the lowest cultural deposits span a time frame of *517 calibrated years*, followed by an approximately *6000-year gap* from the higher levels, suggesting a long

Table 8.1 Radiocarbon dating of charcoal samples results

Excavation level	Depth from surface	^{14}C sample weight	Raw date BP	Range ± years	Cal year BP (95%) IntCal 20	
					Maximum	Minimum
2 (objet #1)	~18 cm	1.1 mg	1180	±30	1178	957
4 (object #12)	60 cm	0.5 mg	3740	±30	4216	3920
5 (object #20)	72 cm	1.6 mg	5660	±30	6486	6310
6 (object #17)	87 cm	1.6 mg	11,212	69	13,236	12,927
7 (object #2)	90 cm	0.4 mg	10,993	62	13,069	12,755
7 (object #13)	82 cm	1.43 mg	11,088	64	13,100	12,790
7 (object #14)	87 cm	1.97 mg	11,148	71	13,170	12,850
8 bedrock (object #5)	108 cm	0.07 mg	>30,300	NA	>34,648	

uninhabited period that correlates to the "*sterile*" layers reported by the Roosevelt et al. (1996) excavation at the nearby cave, *Caverna da Pedra Pintada.*

Charcoal found near the very bottom of this excavation pit (*bedrock*) at *108 cm* yielded a date of >*30,300 uncalibrated years BP*. It is possible that this sample was too depleted of ^{14}C to give a more accurate reading.

At the very least, the date can serve as the baseline sterile period of soil accretion for the entire site (Table 8.1).

Lithics Data

Of the 1688 lithic pieces collected from the 1.4-meter depth of sandy loess soil in the ~1.6-meter width × ~1.8-meter length excavation (Fig. 8.5), 110 artifact samples were analyzed, rinsed, then analyzed again microscopically (×20 to ×30 magnification) to determine evidence of knapping, retouch, use-wear damage, and surface treatments or alterations (including polishing, painting, demarcating, and flame exposure). All the specimens were discovered in situ, as distinct objects (recorded as artifact *object* numbers) during the excavation. Not microscopically analyzed were another 524 lithics recovered in *clusters* from distinctly confined regions of the pit. The *clustered* lithics were so labeled and recognized as resembling cursory knapping flakes and debitage. An additional 126 lithic artifacts that were also discovered in various groups shared distinct contextual *features* (style, coloration, and morphology), and were thus labeled for their distinct appearances.

One such *feature* lithic was previously published (Davis et al. 2017)—a rounded boulder with red and yellow paint (Fig. 8.6). This boulder, with clearly visible red and yellow paint, even after being buried for ~13,000 years, was next to one of the dated charcoal samples (object #2) from level 7, firmly establishing that time frame as a period of painting activity for the site. The rest of the 918 out of 1688 lithic pieces were either recovered during screening or were cobbles and small boulders encountered during excavation. They appeared to be natural rock ecofacts (or chipped off the shelter walls) with no discernible cultural significance.

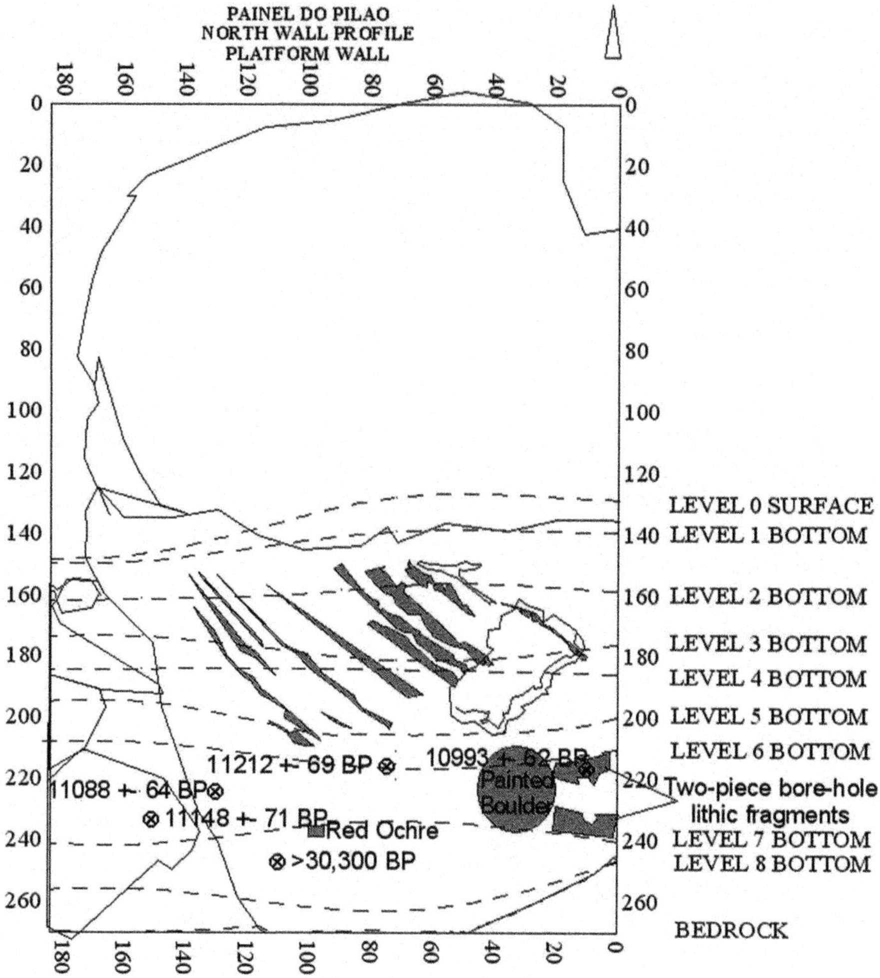

Fig. 8.5 Excavation profile showing the platform wall and the excavation levels, the radiocarbon ages of charcoal from their stratigraphic locations, and art-associated artifacts. The size and irregularity of boulders in the lower stratigraphic levels belies the depositional sequence of features. The resting base of the three artifacts (red ochre manuport, painted boulder, and the oldest fragment of the two-piece polished bore-hole fragments) lie underneath all the relevant charcoal dates

Several additional boulders too heavy to be fully removed from the pit were not counted at all as lithics but were labeled alphabetically from A to Z as encountered at depth during excavation. Some of these large boulders appear to have been altered and flaked, particularly boulder X, Y, Z (Fig. 8.7), which was assigned three letters because three different protrusions of it were unearthed before they were revealed to be part of a single boulder. This boulder has a large central scoop mark which may have been the "scar" of a prior portion flaked off before this boulder itself. Although potential human alterations of the boulders are culturally significant, they

Fig. 8.6 Red and yellow stained circular rock (feature 8) excavated from level 7 at *Painel do Pilão*. D-stretch_lab enhanced (above), raw photo (below)

were all too massive to be taken back to a lab for analysis. However, their presence should be taken as further evidence for knapping activity, and perhaps site function as a sandstone quarry. Many other large boulders appeared to have been unaltered, except perhaps when considering how they came to rest in the shelter space very long ago—from natural wall fall, or from intentional detachment.

Fig. 8.7 Excavation pit photo of Level 7 Boulder XYZ below the color scale. This overhead view of the excavation pit depicts the triangular boulder with an exposed concave surface that still exhibits moisture stains (from the recently removed soil)

Of the *110 microscopically analyzed object artifacts*, *31 show signs of use* against a hard surface like wood or stone. This alone is not inconsistent with the common perception of stone-age cultures, but what is informative is the way the shelter space was used and transformed, which indicates either *lithic sourcing or modest terraforming*, but without complete debris clearing. The lingering mystery is why *one wall* was given significant time and labor to flatten while the other two seem mostly unaltered.

Additionally, *43 of the analyzed artifacts* show evidence for *artistry* based on their additional surface treatments, this group is referred to as *artisanal lithics*. Some of these have *pigmented surfaces* (*25 samples*) due to contact with (or perhaps were themselves directly painted with) pigments matching the *red and yellow hues* still visible on the overhanging sheer rock wall. Other lithics were *polished* (*8 samples*) or favor a *decorative* (*7 samples*) rather than utilitarian appearance; *two objects* show signs of *heat treatment*.

Several lithics displayed dark lines (13 samples) on their surface, either from now-decayed cordage perhaps used for hafting, or from intentional demarcation for reasons unknown (see Fig. 8.8). These dark lines were most visible when they were first unearthed. It is not caused from dirt that caked up along cracks in the rocks, the lines are all on flush surfaces. The left image shows a large lithic with a curved dark line and three shorter lines branching horizontally, almost like an F, because the longest branch is at the top. The middle image displays a bold dark diagonal line near the top of a lithic wedge above a red stain on the beige-colored rock. The image

Fig. 8.8 Lithics excavated from level 8, with dark linear features possibly from decayed lashings, or intentional demarcation. All three images are from video still frames taken during excavation

on the right side shows lithics discovered in a group, and all noticeably with red marks. The largest lithic near the top has dark lines in a Y or broken X shape, across the dorsal flake-scar ridge. Some of the 43 artisanal artifacts have more than one surface alteration. The remaining 36 *object* lithics are characteristic of blanks or preforms.

The *object* lithics were categorized by perceived functionality based on shape and microscopic detail. Two main divisions in analysis emerged: utilitarian stone or woodwork features, and artisanal features. Utilitarian features were subdivided into wedges (thick triangular shapes possibly used to pry or be hammered to split wood or rock open), axe/adze (probably hafted and swung to either cleave or chip wood or rock), scraper (used to shave or level wood, rock, or organic materials), polisher/abrader (to grind or rub a wood or rock surface into a smooth finish), drill/chisel (to peck or gouge out depressions, holes, or indentations). Evidence of organic residue was slight and generally did not fit expectations of food processing. Instead, it perhaps came from mastic substances, like resin, sap, clay, egg yolk, or possibly blood. When discernible, wood (w) or stone (s) was further guessed as the medium worked by the tool, but for many objects, microscopic detail was insufficient for discernment—although the general sense is that wood was more recognizable due to smudging, while stone was the default medium if there was contact polishing (see Table 8.2). Regardless, the medium worked should not be taken as more than guesses, at best.

Lastly, the initial shelter space may have served as a small quarry, one of just a few described this early in the Americas. Most of the retrieved lithic chunks reflect varied tool forms, even if incomplete or cursory. A few of the boulders that were too large to be taken for analysis nonetheless had signs that flakes had been knocked off of them in antiquity, but the pit had relatively little lithic micro debitage—suggesting many detached rocks were possibly taken elsewhere to be more formally knapped. However, the overall data sequence at *Painel do Pilão* suggests that Paleoindians conceived of the space in manifold and subsequent ways: potentially as a sandstone quarry, a shelter (Davis 2009), a rock art platform base (Davis 2014), and a horizon-sighting outpost.

Table 8.2 Stone, wood, or artisanal style of lithic artifacts

		Stone or wood work tool (w_wood/s_stone)					Artisanry/Ornamental										
Level	Object #	Wedge	axe/ adze	Scraper	Polisher	Drill/ chisel	Decorative	Polished	Red pigment	Yellow	Orangish	Purple	Dark Lines	Mud/ slip	Heat exposed	Weight (g)	Pieces
1	1															1422	1
1	2	1														341	1
1	3		1													92	1
1	–															41	7
3	1	1														45	3
3	3		1													563	1
3	4															402	4
3	5		1													480	2
3	6															82	1
3	7															31	1
3	8				1			1								81	1
3	9															65	1
3	10															295	1
3	11															58	1
3	12								3	1						66	3
3	13															66	1
3	14															86	1
3	15															88	1
3	16															39	1
3	17			1w												65	1
3	18															599	1
3	19							1								43	1
3	20		1	1w												151	1

3	21				1w			1								62	1
3	22		1						1							319	1
3	23															52	1
3	24															118	6
3	26			1w												80	1
4	5															336	1
4	6			1s												741	1
4	9															1069	1
4	10															248	2
4	11		1													302	1
5	1															524	1
5	2	1														683	1
5	3		1											1		260	1
5	4													1		326	1
5	5															249	1
5	6															433	1
5	7															120	1
5	8	1														402	1
5	9												1			2100	1
5	10															109	2
5	11															51	1
5	12				1s								1			110	1

(continued)

Table 8.2 (continued)

Level	Object #	Stone or wood work tool (w_wood/s_stone)					Artisanry/Ornamental									Weight (g)	Pieces
		Wedge	axe/ adze	Scraper	Polisher	Drill/ chisel	Decorative	Polished	Red pigment	Yellow	Orangish	Purple	Dark Lines	Mud/ slip	Heat exposed		
5	13								1	1						151	1
5	14												1			510	1
5	15															9	1
5	16		1													154	1
5	17		1										1			341	1
5	18	1				1s								1		122	1
5	21															86	1
6	1	1						1								134	1
6	2	1						1								132	1
6	3		1													88	1
6	5															76	1
6	6					1s										57	1
6	7															1240	1
6	9								1	1						1000	1
6	10		1													181	1
6	11							1						1	1	420	1
6	13		1					1						1		380	2
6	14															125	1
6	16															43	1
7	1						1									1530	1
7	3										1					1100	1
7	4															14	1
7	5								1	1	1					15	3

7	6										1					74	1
7	7						1									16	1
7	8		1													63	1
7	9					1s							1			1233	1
7	11															23	1
7	12												1			176	1
7	15												1		1	90	1
7	16						1						1			2350	1
7	18						2	1	1				2			855	2
7	19								1							154	1
8	1		1	1w				1			1					215	1
8	2			1w									1			1202	1
8	4			1w							2					775	2
8	6								1							75	1
8	7										1					65	2
8	8												1			782	1
8	10												1			743	1
8	11						1	1								20	2
8	12						1	1				1				80	3
3b	–								1							171	3
Tally:		7	14	7	1	0	7	11	11	4	7	1	13	5	2	29,069	110

Stone and Woodwork Evidence

Level 3 of the excavation did not yield charcoal directly radiocarbon dated, but 24 *object* artifacts come from this level and 19 were categorized with functionality. Four among those artifacts have use-wear damage possibly analogous to woodwork. Level 4 had far fewer artifacts than those below or above it. Level 5 of the excavation increased in lithics and features suggesting the intention for stone or wood use, including a scraper (Fig. 8.9), a drill, and wedge, axe, or adze-like forms (Fig. 8.10).

Fig. 8.9 Sandstone flake from level 5 object 14, resembling an axe. The (**a**) dorsal side is highly polished, the (**b**) lateral edge (where it may have been intended for hafting) is tapered into an acute triangular edge, and the (**c**) ventral side shows shallow, pitted regions and indentations. Under ×20 magnification, the (**d**) ventral edge shows use-wear polish, the (**e**) lateral edge shows polish and slight whitish residue, the (**f**) magnified dorsal surface supports the claim of heavy, intentional polish, and (**g**) is the magnified blueish region, showing the coloration is due to the deterioration of surface polish to reveal dirt embedded in coarse sand crystal

There are a few lithics with shapes and micro-wear evidence suggestive of more conventional subsistence-like cutting use. One of the few lithic artifacts that reflect this include a tetrahedral artifact from level 6, which may have been used as a perforator, burin, or knife (Fig. 8.11), and a triangular flake which may have been used as a projectile point or knife (Fig. 8.12). There are darker stains or smudges just at the tip, which may have been due to cutting or piercing animals or plants.

Further wood or stonework evidence was found in Level 6, including a small axe-shaped stone tool (Fig. 8.13). This artifact has a highly polished dorsal surface

Fig. 8.10 Sandstone wedge from level 5, object 17. The (**a**) dorsal side is highly polished, with two transverse, opposing flake scars forming a midline ridge. The (**b**) ventral side is concave. Under ×20 magnification, the steep-angled distal end (**c**) has evidence of polish. The lateral edge of the steep front has (**d**) dark use-wear evidence on top of a whitish coating. The ventral-side distal tip shows further signs of reddish residue and contact polish from scraping or other impact use-wear damage

Fig. 8.11 Elongated tetrahedron-shaped flake (object 14) from level 6 displaying the (**a**) lighter dorsal side, the (**b**) jagged dorsal side, the (**c**) ventral side with the distal tip to the left, and the (**d**) ventral side with the distal tip to the right. At ×20 magnification, the (**e**) distal tip shows use-wear damage and a pale "crust." On the mid-dorsal ridge is a (**f**) flat and polished region. The general dorsal surface also has residual evidence of a (**g**) whitish coating, and the magnified (**h**) proximal end shows a well-crafted acute angle, worn down on its edges

except for the back region, where it tapers down to its narrowest edge. Level 6, and lower, all date to the earliest Paleoindian period before the Younger Dryas.

A larger stone adze-like tool was also discovered intriguingly alongside a burin-like tool (Fig. 8.14). This pair was excavated together, and appear to have been deposited together, in situ, long ago. Low-level magnification (20×) of this "adze" reveals apparent evidence that wear on the edge is not found on the bulk of the tool, despite having been buried, so the visible edge abrasions reasonably come from intentional tool use. The "adze" also appears to be multifunctional, with the distal end truncated to a broad scraper-like edge while the proximal end has been laterally tapered to an axe-like edge. The small flake found alongside it has a hardened spur, perhaps utilized as a burin-like engraver. Seen under 20× magnification, the spur edge appears to be encrusted with an off-white coating as though it was dragged through softer, mud-like material. Considering the context in which they were discovered, both adze and burin may have been used to modify the wall, or some other unknown substrate.

Massive adze-like flakes suitable to plane wood or stone surfaces were also unearthed in level 7 (Fig. 8.15) and level 8 (Fig. 8.16). Several lithics recovered from level 8 were also wedge shaped, with low magnification showing subtle signs of usage (Fig. 8.17). The shape and cursory use-wear evidence suggests lithic knapping for uses related to leveling or flattening wood or stone. However, the presence of surface coloration (typically only on one side) of an off-white coating suggests factors other than simple utilitarian or impulsive use had occurred. Unifacial polishing of level 8 lithics (Figs. 8.18 and 8.19), and an afore-mentioned off-white coating of several—but not all—lithics from levels 5 through 7 provide further evidence discussed in greater detail in the artisanal section below.

Fig. 8.12 Triangular flake (object 16) from level 6 displaying the (**a**) ventral side, where there is a lengthwise upraised ridge near the right edge. The center of the flake is flatter than the surrounding areas, with the ridge suggestive that the flake may have been hafted. The (**b**) dorsal side depth details the left edge is thicker, as is the tip, while the back edge is tapered. At ×30 magnification, the (**c**) dorsal-side tip shows use-wear smudging, and the (**d**) dorsal-side mid surface shows signs of polish. The sum of evidence suggests this tool may have been hafted and used as a projectile point or a knife on the "serrated" edge

Fig. 8.13 Small axe-like wedge from level 6 object 1. The (**a**) dorsal side and the (**b**) ventral side are polished over most of their surfaces when viewed under (**e**) ×20 magnification. The (**c**) lateral left side is depicted. The narrowest edge is the (**d**) distal, chipped, and axe-like edge

Fig. 8.14 Large wedge flake (adze?) and small flake (burin) found together from level 6 object 13. The large flake (**a**) dorsal side, (**b**) ventral side, (**c**) lateral right-side, and (**d**) lateral left side. And at 20× magnification: (**e**) retouch scar of distal edge, (**f**) use scar, (**g**) dulled utilized cleaving edge at proximal end, (**h**) utilized encrusted spur of burin flake

Certain regions in level 7 contained lithic clusters indicating possible activity areas, like the M-region microliths (Fig. 8.20). G-region lithics also represented chipped flakes recovered near—and possibly flaked from—the east/northeast wall near boulders *E* and *F* in level 7. Feature 5 lithics (resembling large pecking tools like chisels) were similarly recovered directly adjacent to the wall, as well as a massive flake (Fig. 8.21) that may have been directly detached from the wall.

As mentioned above, the rest of the *object* lithics (36 specimens) mostly fit the description of preforms or quarry blanks (Funk et al. 1969:5)—relatively thick

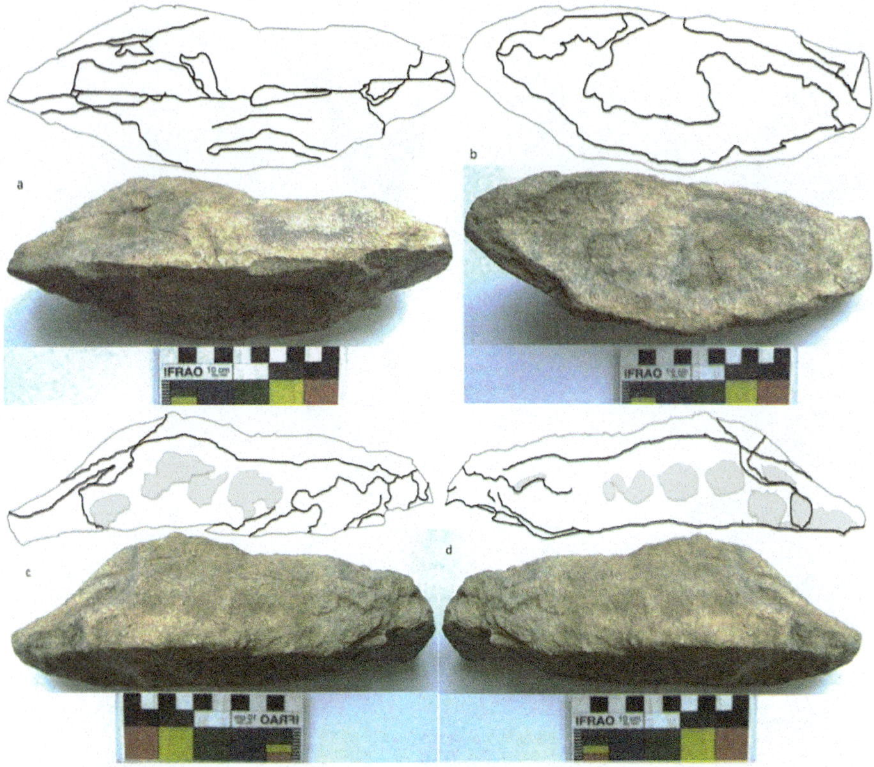

Fig. 8.15 Heavy plano-convex flake from level 7 object 9. The (**a**) dorsal side. The (**b**) ventral-side flat bottom could have been a scraper/sander, with ends used perhaps as a hammer or pile driver. The gray polygons on the sketches of both the lateral left (**c**) and lateral right (**d**) sides are indentations spaced consistent with utilized finger grip positions

flakes with modest signs of percussion chipping or grinding reduction from a core or a spall, and virtually no sign of retouch. However, these stones are just as significant because they prove that the claims of human modification mentioned above (whether artistry or wear patterns) were not produced naturally or incidentally by varnish, accretion, water seepage, or so on.

The modifications mentioned above are not found on these "blanks" nor are they found on the excavation boulders.

In conclusion, stone and woodwork wear patterns are most evident from low-grade microscopic analysis of *object* artifacts from levels five to eight, with the most massive artifacts unearthed from levels six to eight—the earliest cultural periods deposited at this site. Some of the level 5 objects (which date to ~6400 years ago) with similar features and characteristics to the Paleoindian-period lithics may very well have been some of these older artifacts rediscovered as surface, or near-surface, finds during the pre-pottery Archaic period after ~6000 years of sterility. While many artifacts appear to have been tools or preforms, some lithics may have been detached from the shelter wall, providing supporting evidence that suggests stonework was performed to alter the shelter space itself.

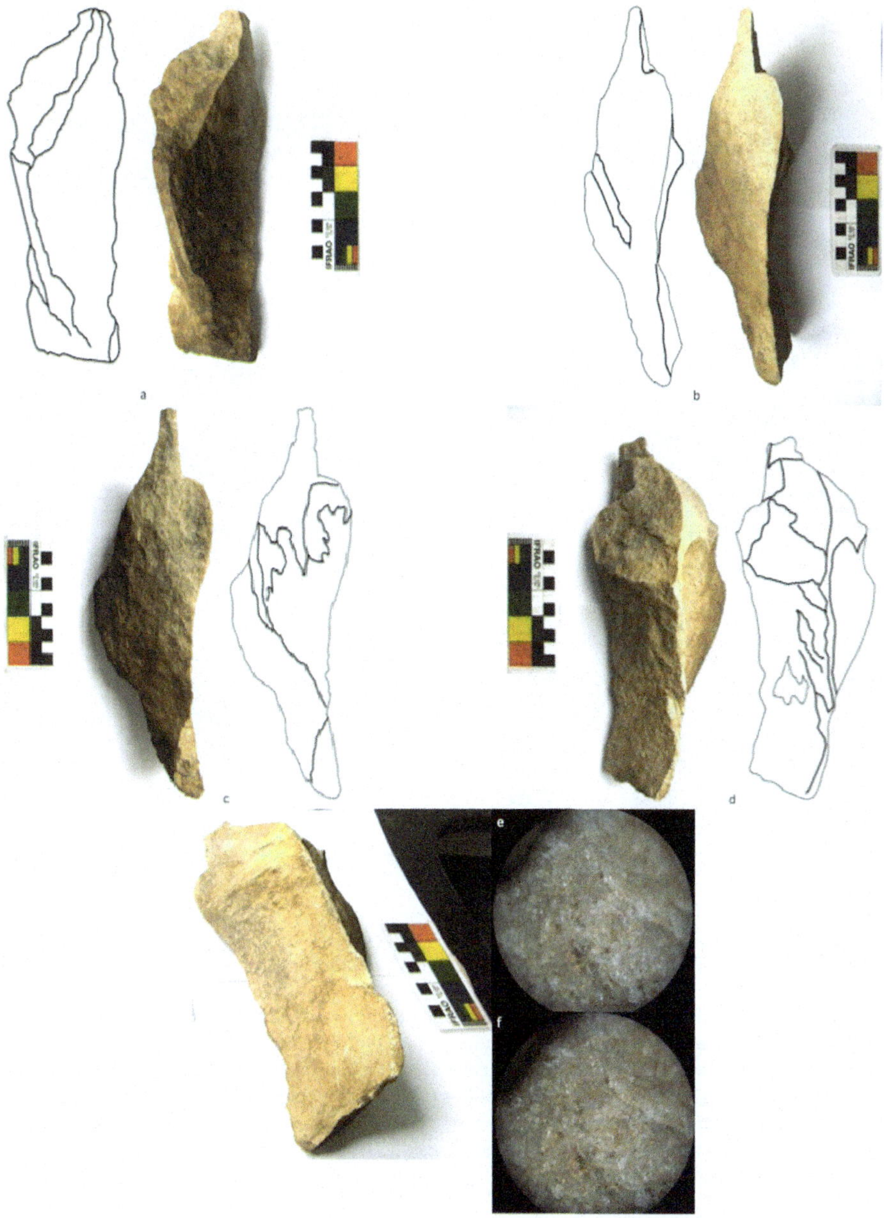

Fig. 8.16 Long humped-back flake from level 8 object 2. This massive flake has a humpbacked spine visible from the (**a**) dorsal view, a contracting stem at the proximal end most visible from the (**b**) lateral right view, which also shows a large smudge stain and evidence of a mid-range flat region (flute?). (**c**) The lateral left view shows the truncated ovoid distal end useful for end scraper activity. (**d**) The ventral side shows a curved broad bottom surface and scraper-like edge at the distal end, plus a deep curved indentation on the left side of the humped-back handle. (**e**, **f**) show 20× magnification of a portion of the ventral side, polished possibly intentionally done or unintentionally due to wear patterns from leveling wood

Fig. 8.17 Wedge-shaped flake from level 8 object 7 contours to the right hand like an engraver. (**a**, **c**) Dorsal side showing shallow perpendicular mid-dorsal flat flake scar and possible polish. (**b**) Ventral side showing dark smudges from use. (**d**) ×20 magnification of distal tip indicates slight dark smudge and polish wear patterns, possibly from cutting/engraving. (**e**) ×20 magnification of mid-ventral parallel ridge indicates dark (organic?) residue on edge. (**f**) ×20 magnification of left lateral edge indicating similar reddish smudge residue from cutting or scraping, and some slight polish, possibly also from use

Evidence for Applied Pigment and Artistry

Some lithics with apparently deliberate pigment were excavated in the lowest levels, along with chunks of pigment ore—like ochre. Beyond the red and yellow hues similarly found in the pictographs from the previously mentioned rock art wall, several flakes had white or off-white to cream-colored surface coatings. One lithic from level 3, a much more recent layer, is unmistakably painted, with an evident

Fig. 8.18 Round-edged, wedge-shaped, large flake from level 8 object 8. (**a**) Ventral side showing dark thin (lashing?) lines at top, and two stepwise ridges on the right edge, one of which is under (**c**) ×20 magnification. The dorsal side (**b**) shows a shallow dented, and somewhat polished, surface (**d**) displayed at ×20 magnification. The back edge of the wedge is flat enough that the flake can stand upright (**e**) with the dorsal side facing toward the top of the image and the tip seemingly broken. With the (**f**) dorsal side facing the bottom of the image, the ventral surface contours are more easily visible

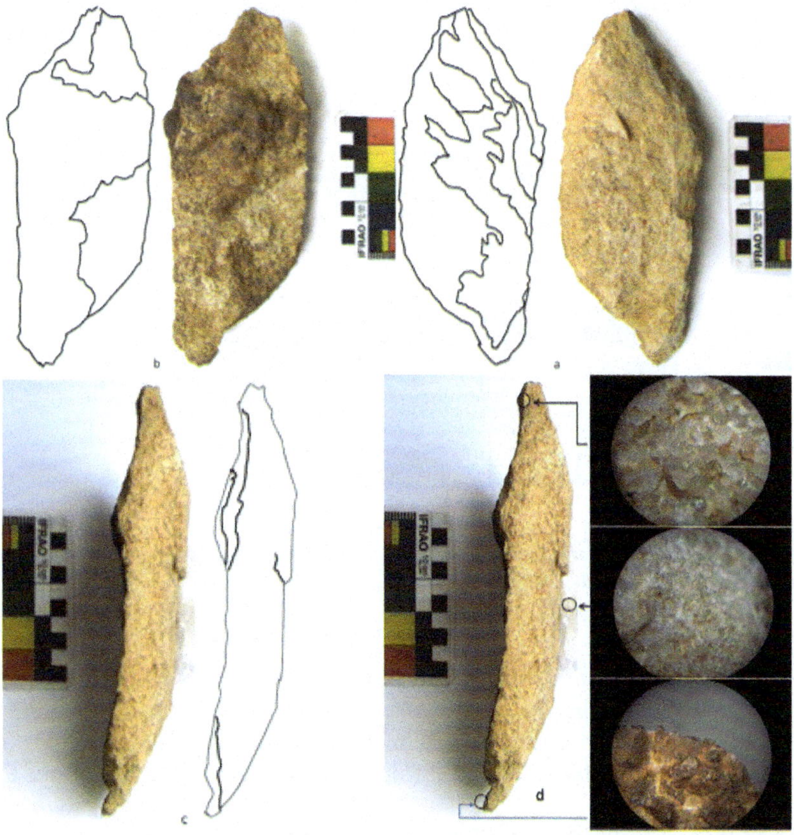

Fig. 8.19 Long flake from level 8 object 10. (**a**) Dorsal side showing shallow flake scars and possible polish (**b**) ventral side showing dark smudges from use, perhaps scraping (**c**) lateral left side reveals how flat the mid-dorsal region is, perhaps from scraping wear patterns instead of intentional fluting (**d**) ×20 magnification at three different positions: the proximal dorsal edge (tang), the polished-like flat mid-dorsal region, and the ventral distal tip showing dark residue

Fig. 8.20 Some of the M-region microliths and flaked "chips" from a localized concentration in level 7, suggesting that the area was associated with a particular knapping activity or perhaps removed from the east/northeast wall

stroke of red pigment on its white-coated dorsal region (Fig. 8.22). It is possible that some of the painted lithics were painted shelter wall sections that were later detached, but the pattern and fracture of most of the lithics do not support the theory that most "decorated" flakes occurred from this fashion, nor that they broke off the wall naturally.

One lithic artifact that is clearly painted in both red and yellow color (and clearly not a wall fragment) comes from level 5. It is a large, relatively narrow, flat stone that is unusually sturdy to be sandstone—perhaps due to higher iron content. If it is a natural mineral, it was no doubt a manuport carried and deposited in the *Painel do*

Fig. 8.21 Massive flake from level 7 object 3. Over 15 cm long, the flake was discovered adjacent to the east/northeast wall. It is large enough to have been quarried directly from the wall. (**a**) Dorsal side has percussive flake scars, and the top right corner has a smudge mark, (**b**) ventral side has no recognizably significant marks

Fig. 8.22 Large sandstone bulb from level 3 object 22. The (**a**) dorsal side has a prominent red stroke on a region that was painted white. The (**b**) ventral side has a lengthwise, midline ridge. Under 30× magnification, (**c**) the boundary of red and white paint can be seen, with parts of the red area revealing vertical streaks that indicate it was layered on top of the white paint. Compared to the (**d**) distal end of the dorsal side, where there is no paint, but perhaps some either intentional or contact polish. And (**e**) magnifies slight wear patterns on the ventral side ridge

Pilão shelter (Fig. 8.23). The *exterior or dorsal side* of the artifact is identified as the side with *5 or more shallow narrow grooves* on a flanged *"rim."* The numerous *narrow grooves* on one edge give it a resemblance of a *stone vessel border*, but it is uncertain if this artifact is indeed a fragment from a vessel. The *interior or ventral side* most distinctly shows that half of its surface has *yellow pigment*, while the other half has *red pigment*—visible under ×*20 magnification*. Under the same magnification level, the *exterior* also shows *red pigment*, although much less. A broken section of the rim shows that the *red color does not continue* all the way through the

Fig. 8.23 Large flat artifact from level 5 object 13—(**a**) Exterior or dorsal side with 5 or more shallow narrow grooves on a "flanged rim." (**b**) Interior or ventral side under ×20 magnification shows half its surface has (**c**) yellow pigment, half has (**d**) red pigment. Exterior surface magnification shows (**e**) less concentrated red pigment. A broken section of the rim (**f**) shows that the red color is just a surface coating. Grooves on the rim are (**g**) deeper here, while (**h**) shallower here. The exterior side appears to have a different pattern of (**h**) yellow, (**i**) orange (perhaps where yellow and red pigment overlap), and (**j**) red region distributed differently than the interior side

material substrate, but instead is just a *surface coating*. Some of the *grooves on the rim* are deeper, while others are shallower, and *yellow paint* appears on the *exterior side* as well. Some spots appear to be *orange*, perhaps where *yellow and red pigment overlapped each other*.

Other artistic features include shaped lithics in seemingly non-utilitarian styles. A sandstone block, almost in the shape of a rectangular brick from level 7 (Fig. 8.24)

Fig. 8.24 Rectangular sandstone "brick" (object 1) from level 7 displays two notches on side (**a**) where a dark rectangular notch is visible at the top, and a squarish notch is at the bottom, on the front-facing side. Side (**b**) has a shallow u-like flake scar 1/third of the way up from the bottom. The square notch on side a is now to the right. Side (**c**) is slightly bulged convex, and the other side of the dark top notch is seen at the top. Side (**d**) seems the flattest side, and the side c bulge now faces to the right. Side (**e**) is a top-down view of that dark topnotch, where now it is noticeable that one side is triangular. Side (**f**) is the "bottom" of the "brick." The two notches were probably altered by humans

was one of two brick shapes excavated. Another large sandstone artifact excavated from level 7 (Fig. 8.25) could be considered an engraving due to its deep narrow grooves, two on the side, one at the top, and a few faint grooves on the top-right. The overall shape resembles a right hand or paw, although no specific conclusion can be drawn.

Another more notable decorated lithic includes a unifacial, talon-shaped flake from level 8 (Fig. 8.26). The oblique lateral right side (distal end) appears to be the cutting or scraping edge, and the piece has a triangular-shaped proximal end. Its dorsal side exhibits signs of polishing and a mid-range flat region (flute?) while the ventral side shows no evident retouch. Under 20× magnification, the distal edge shows some wear as though it was used for scraping, while the dorsal left lateral edge shows evidence of blunting due to use, but also it exhibits a white surface coating in contrast to the ventral left lateral edge. Magnification also enlarged a dorsal "spine" flake scar above the flute, which has faint red stains on the right side. This piece is not the only lithic that has a mid-range flat region (flute?), nor is it the only lithic with a whitish surface coating, indicating that modifications went beyond simple utilitarian or impulsive preparation.

Fig. 8.25 Grooved sandstone block (object 16) from level 7 of the (**a**) dorsal and (**b**) ventral side are presented as (left) unwashed to show the dark lines as possibly decayed lashings, and (right) washed surfaces for magnification. The ×30 magnification of the dorsal side (**c**), near the top grove shows very fine polish giving the appearance as though crystalline sand grains are fused together. Magnification of the ventral side (**d**) (palm?) shows a whitish crust, or perhaps a crystalline precipitate that formed on the surface, long after disuse. The lateral right side (**e**) shows the cleft that resembles the thumb of the piece, while the lateral left side (**f**) shows the straight edge. The in-situ excavation photo (**g**), along with the unwashed dorsal side (**h**) and the unwashed ventral side (**i**) show faint, thin dark lines across its surface, possibly from long decayed lashings

Fig. 8.26 Unifacial talon-like flake from level 8 object 1. (**a**) Dorsal side with signs of polishing and a mid-range flute. (**b**) Oblique lateral right side fronting the cutting or scraping edge and triangular-shaped proximal end. (**c**) Ventral side showing no evident retouch. (**d**) Lateral right profile, (**e**) lateral left profile, (**f**) distal end, (**g**) proximal end, and (**h**) 20× magnification of dorsal spine flake scar above flute, showing some red stains on the right side, (**i**) 20× magnification of distal end edge, showing some wear as though its end was used for scraping, (**j**) 20× magnification of dorsal side lateral left edge shows evidence of blunting due to use, but also a white surface coating in contrast to (**k**) 20× magnification of ventral side lateral left edge

A lanceolate-like (or falcate) flake (Fig. 8.27) was recovered from level 5. It has no apparent evidence of alteration to its ventral side either, except for an apparent white coating (not cortex or patina) on parts of both the dorsal and ventral sides. The off-white coating appears to have been applied as a viscous liquid, perhaps like paint or a clay slip, or perhaps organic (food) matter, and it seems heavier on the

Fig. 8.27 Lanceolate-like incurvate flake with slight shoulder, contracting stem, and concave base, from level 5, object 11. (**a**) Dorsal side showing flat center portion where a flake or flute was removed, (**b**) ventral side seems mostly unaltered, (**c**) broken tip at distal end of flake under 20× magnification displays triangular cross section and an off-white coating, (**d**) edge of contracting stem as seen from the ventral side under 20× magnification (**e**) lateral left side view, (**f**) lateral right side view, (**g**) proximal view, (**h**) distal view

Fig. 8.28 Level 5 object 8: Unifacial flake (probable sidescraper). (**a**) Dorsal side with off-white coating. (**b**) Ventral side shows little to no modifications. (**c**) Lateral view displaying presumed scraping edge. (**d**) 20× magnification of dorsal side of scraping edge shows off-white coating. (**e**) 20× magnification of ventral side scraping edge lacks off-white coating but suggests dark brown staining and dulling of edge due to possible cursory use

ventral left lateral edge. Therefore, it is unlikely that the coating accreted naturally from salt or similar precipitate after deposition.

The lanceolate-like flake is not the only artifact with an apparent off-white coating from level 5. Another side scraping tool appears to have been similarly coated, but just on its dorsal side (see Fig. 8.28), making it more likely an artistic production than resulting from incidental utilitarian use. Unlike the lanceolate flake, the side scraper does have low-magnification microscopic evidence of at least cursory use, and it is one of several scrapers found in this level. A rhomboid flake from this level has soot-like smudges on its dorsal side, suggesting it came in contact with fire or burned wood (see Fig. 8.29).

Of related importance is the fact that lithic flakes with similar unifacial white coatings were discovered in the cache at the Serra do Sol site ~5 km to the north of this site (see Chap. 7). Some of those lithics also had potential fire exposure. X-ray

Fig. 8.29 Level 5 object 16: rhomboid flake (side/end scraper). Photo shows the (**a**) dorsal side with offset median ridge, the (**b**) ventral side seems less altered. The (**c**) lateral view shows probable scraping edge, and the (**d**) blackened median ridge at 20× magnification with the (**e**) distal end ventral side edge at 20× magnification shows use wear; the (**f**) lateral edge at 20× magnification shows heavier use wear

fractionation analysis (pXRF) of the white coating on sample #2 of the cached lithics revealed high levels (7%) of phosphorous. Bird guano or egg yolk is one suspected source, of which local vulture and owl species provide copious white-colored (though foul-smelling) amounts as potential additives to white-to-off-white ochre, clay, or resin.

Additionally, an off-white coating similar in appearance also lies underneath many of the rock art paintings, like a backdrop, on the vertical walls at Painel do Pilão and other painted walls in the region. This suggests to me that the painted walls were first "primed" with a pale backdrop before red and yellow drawings were painted on them. Regardless of the reasons or beliefs they may have held—which we can only guess and imagine—this consistency from lithics to walls may be considered a cultural "style" for the Paleoindian period; a cognitive or symbolically understood way of doing things. Adding further evidence to this association is the raw red ochre chunk found in the lowest Paleoindian excavation levels (Fig. 8.30).

Excavation level 6 revealed tools with direct evidence for directly processing pigment and some of the rocks that may have been the direct source of pigment. This stone "tablet" may have been used akin to a palette, or perhaps was itself used as a source for red ochre pigment (Fig. 8.31). At the time that it was excavated, it was covered in superfine dark dust (soot), which may have been due to exposure to

Fig. 8.30 Raw reddish-orange ochre found in level 8 (object 6) of *Painel do Pilão* excavation.

Fig. 8.31 Thick rectangular red "tablet" (object 1) from level 6. The unwashed (**A**) dorsal and (**B**) ventral sides display the dark (soot-like) coating of the rock when it was first excavated compared to the washed (*a*) dorsal side and (*b*) ventral sides. The four lateral sides are (*c*) lateral step side, (*d*) lateral straight side, (*e*) lateral nodule side, and (*f*) lateral red side

fire. If so, fire may have been an important step in pigment production from ochre-rich clay or rock.

One pointed tool has red pigment stains transferred perpendicular to its distal endpoint (Fig. 8.32). One side is thicker, with ridges that suggest it may have been hafted and swung downward like a pickaxe. The orientation of the reddish-brown marks suggests the tool was used at least once to pierce a large reddish ochre nodule, or some other pigmented object. The former scenario is supported by the fact that a few small pigment manuports were recovered in levels 6, 7 (Fig. 8.33), and 8. No pigment nodules were discovered in direct surface-to-surface contact with any stained lithics, and since these red marks were not removed during the rinsing preparation for analysis, incidental contact coloration of any artifact is unsupported by in situ excavation evidence.

A lithic tool from level 7 also has significant paint (Fig. 8.34) in the form of a broad red streak of pigment across its dorsal and ventral surfaces. The D-stretch color enhanced image of it suggests the specimen itself may have been coated in yellow (instead of white) over its entire surface. The red streak here seems more indicative that the tool was used as an end scraper against red pigment, or painted material.

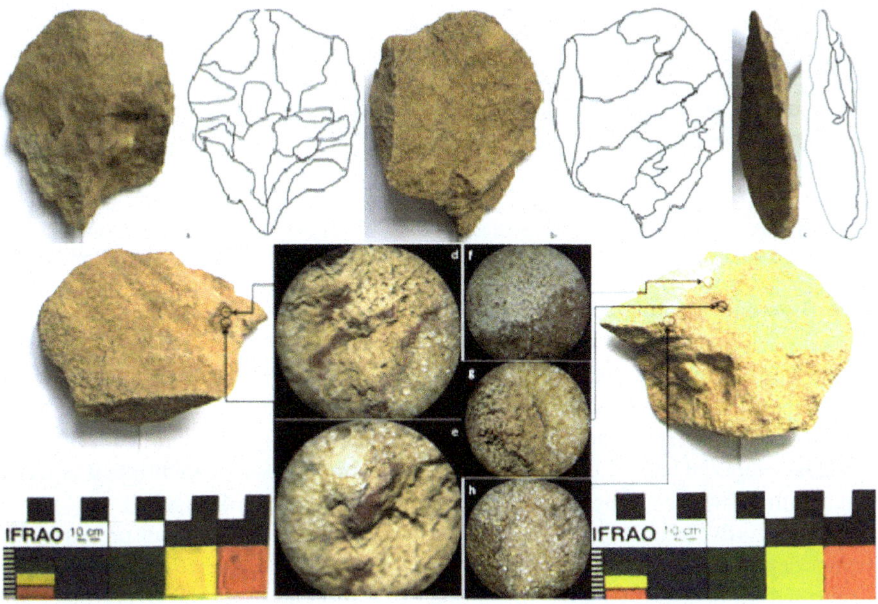

Fig. 8.32 Sandstone cleaver from level 6—object 10. The dorsal side (**a**), ventral side (**b**), lateral view (**c**) brick red pigment at 30× magnification of ventral side near piercing/cleaving edge (**d**, **e**) show red lines on protruding surfaces perpendicular to the perceived direction of cleaving. A dorsal side flake scar with off-white coating (**f**) and the interface boundary of the flake scar (**g**) suggests significant polish (perhaps some from use-wear damage). The most reflective polish appears at the apex of dorsal side (**h**)

Fig. 8.33 Three red and yellow pigment-rich rocks from level 7 object 5, (**a**, **b**) are probable cemented ochre-rich clay and sand (**c**) is probably goethite heated or scorched for greater ease in pigment processing. The right-side image is also from level 7, among the XY region of small artifacts. After washing dirt off in the laboratory, it is a small whitish chalk-like nodule. Perhaps a source for white coating

Fig. 8.34 End scraper used for red paint processing from level 7—object 9. The dorsal side (**a**) has significant red pigment smears on its surface near the distal end, made vivid with Dstretch-LRD image enhancement (**b**). The ventral side (**c**) has a particular red pigment streak from distal to proximal end, made vivid with Dstretch-LRD (**d**). The lateral side (**e**) appears particularly yellow in the Dstretch-LRD image of its right side (**f**)

Evidence of painting activity increased in levels 7 and 8. Level 8 also contained a number of lithics with highly polished dorsal surfaces but less apparent modification of ventral surfaces, including a highly polished, unifacial, triangular point with a concave tapered base (Fig. 8.35). There is also the previously mentioned unifacial flake with off-white coating and faint red surface stains (see Fig. 8.26).

Additionally, a large bored-hole flake (Fig. 8.36) broke into two pieces before being deposited in antiquity. The two fragments of the bored-hole flake were unearthed in the same vicinity within the excavation unit, but the larger piece was found near the base of the level while the smaller fragment was found near the top of the level. The smaller piece has additional evidence of potential reuse as a sidescraper based on its smudge stains and tapered edge. This is an important clue because if the smaller fragment were used for several months, years, or even generations later, it would be unlikely to be deposited so near to the larger fragment. However, if the infill of level 7 occurred anywhere from a day to a few weeks, and the small fragment were used in that span, the tool user may have decided to discard

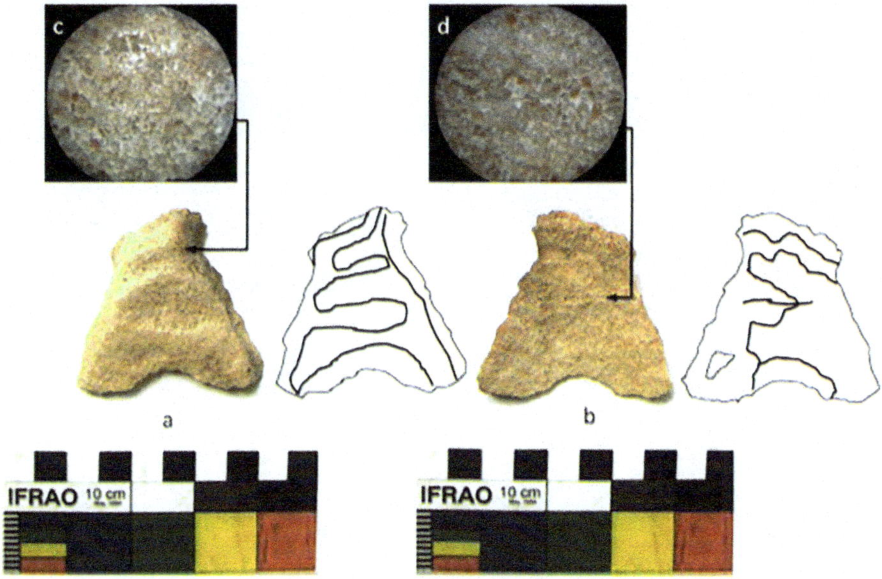

Fig. 8.35 Thin nearly triangular flake with concave tapered base from level 8 object 2. (**a**) Dorsal side is highly polished and has ridges, while the (**b**) ventral side appears mostly unintentionally altered. (**c**) Under 20× magnification, the dorsal surface appears to have been polished compared to (**d**) 20× magnification of the ventral side

it near where the larger piece was discarded. The inference can probably never be fully proven, but it supports the theory that a shelter wall was intentionally carved or quarried.

The bored-hole flake and a few small and large boulders from the base of level 8 (virtually bedrock) have stark natural coloration in the rock, which suggests that rocks with natural vibrant colors embedded in them were initially sought and heavily polished (Fig. 8.37), until eventually such rocks became hard to find or extract. If so, it is a provocative clue to why many painted lithics appear next in the stratigraphic sequence. Based on artifacts from levels 5 to 7, this preference for color possibly transitioned into a painting tradition of their tools—as well as on walls (perhaps marking walls where rocks were removed?) in the form of rock art shortly thereafter.

If *Painel do Pilão* changed from a quarry to an artistic monument, then Paleoindians here demonstrated dynamic cultural and technological traits inconsistent with the utilitarian traits commonly portrayed for hunter-gatherers in the Americas at that time. While conventional wisdom suggests that the vibrantly red and yellow rock art paintings were retouched, restored, or elaborated during more recent millennia, few pigment manuports were discovered in higher excavation levels. Therefore, the function of the site during the archaic, formative, and into the modern period seems to have been more as a commemorative rock art site, after its painted themes and meanings may have been lost, retained, or redefined.

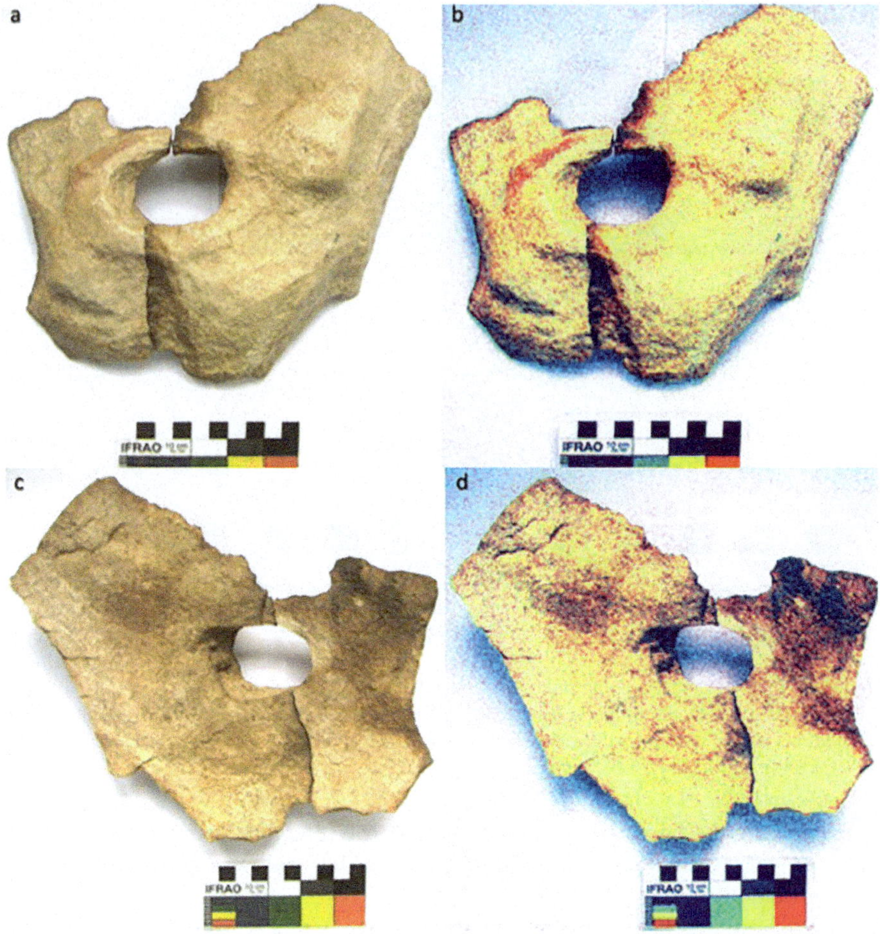

Fig. 8.36 Large bored-hole flake from level 8 object 18. The entire dorsal side of both flakes (**a**) appear to have been polished, while the ventral side (**c**) was not. Dstretch-LRD analysis of the dorsal side (**b**) reveals a streak of red color on the smaller fragment, but this appears to be a natural coloration in the parent rock material. Dstretch-LRD analysis of the ventral side (**d**) shows no artificial or natural color differences beyond the dark edges and smudges due to use, and possibly soot from fire exposure

Fig. 8.37 Purple-mark rounded rectangle from level 8 object 12. The flake, whose right corner broke off during excavation, is (**a**) highly polished on one side but has (**b**) no evidence of polish on the other side. Under 20× magnification, the side that is (**c**) polished has a (**d**) stark purple mark on the surface close to the center. It appears that the purple color is naturally embedded in the rock

Terraformed Wall Data

Compared to the shelter's two other natural walls, which are smooth rocks with long angular edges and multiple vertices, the east/northeast wall has been modified—not in its nascent condition (Figs. 8.38 and 8.39). However, if the wall were simply quarried, we might expect to see multi-planar surfaces at right angles and various depths, indicating repeated attempts to remove rock along more advantageous propagations. Instead, the entire wall was made nearly flat, indicating efforts beyond lithic procurement, toward a goal of terraforming. Contrarily, it seems unnecessary to have flattened this wall simply to serve as a platform. Eleven diagonal gouges and one protrusion described later are the only deviations from the planar surface.

Despite significantly fewer artifacts unearthed in higher stratigraphic levels, the entire ~260-centimeter-high east/northeast wall—above and beneath the initial ground—appears to have been ablated, resulting in a generally flat vertical surface perpendicular to the ground as well as to the platform above. The microsurface shows numerous chips, scoop marks, and flake scars a few centimeters (3 to 5 cm on average) wide. Despite unearthing fewer "chips" than what perhaps should have been recovered, scooped flakes, microliths, and thin or "fragile" flakes associated with knapping from either a wall or from nodules were recovered from the lowest levels. Level 7 yielded 124 "fragile" flakes (thin or fragile enough that it could easily break in your hand), and an additional 74 chip-like flakes recovered while screening the east corner. Level 8 yielded 111 micro-debitage flakes. None of the other levels yielded flakes suggestive of wall-reducing activity (Table 8.3). The presence of thin flakes, but in low overall numbers, might imply that some flakes removed from the wall were collected and taken away, or cleared off from the site.

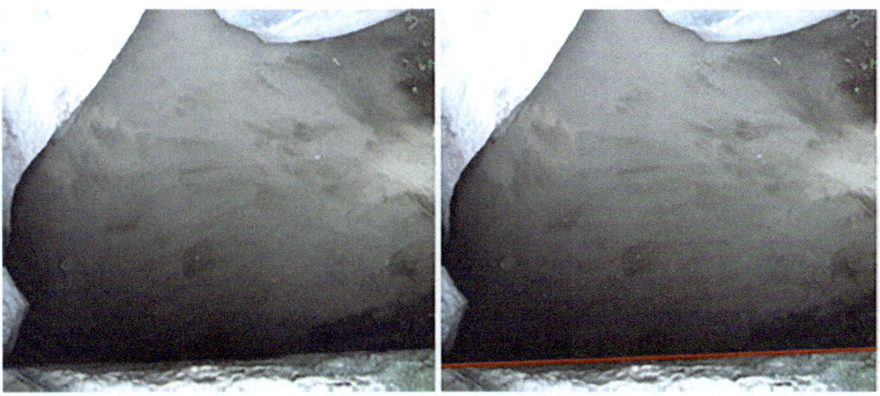

Fig. 8.38 Bird's eye view of the platform wall shows it was quarried and perhaps even terraformed into a straight, flattened surface, when compared to the shelter's two other walls, which are multifaceted, smooth-curved, and arced surfaces

Fig. 8.39 Oblique view of the shelter wall that functions as a platform upon which the rock art can be viewed appears to have been flattened

Table 8.3 "QUARRY" statistics at *Painel do Pilão*

Excavation level	Average depth	Soil volume (m^3)	Lithics by count	Lithics by weight (g)	Weight percent	Percent count	Lithics by volume (avg density ~2.34 g/cm^3)
1	0–16	0.46	10	1896	2.37	0.60	0.0008 m^3
2	16–31	0.43	56	3311	4.14	3.33	0.0014 m^3
3/3b	31–49	0.52	283	6344	7.92	16.85	0.0027 m^3
4	49–62	0.37	127	5208	6.50	7.56	0.0022 m^3
5	62–82	0.58	153	10,436	13.03	9.11	0.0045 m^3
6	82–92	0.29	236	11,919	14.89	14.05	0.0051 m^3
7	92–98	0.17	599	25,145	31.41	35.65	0.0107 m^3
8	98–115	0.49	180	14,297	17.86	10.71	0.0061 m^3
Bedrock (9)	115–137	0.49	36	1506	1.88	2.14	0.0006 m^3
1–5 total	82	2.36	629	27,195	34	37	0.0116 m^3
6–8 total	55	1.58	1051	52867	66	63	0.0226 m^3
Pit total	137	3.94	1680	80,062	100	100	0.0342 m^3
Possible volume of wall removed						180 × 260 × 0.7 cm	

Evidence for Quarrying

In a general sense, a quarry is where subsurface extraction of mineral resources repeatedly occurs for a community; the size, scale, and scope are variable and therefore are non-descriptors. According to R. Michael Stewart (1987), evidence indicative of ancient quarrying activities includes:

> *(A) Scorched sections suggesting fire and water were used to force the rock to spall; (B) Battered outcrop sections representing attempts to detach rock with large hammers, wedges, and chisels; (C) Damaged or discarded quarrying tools deposited in related contexts; (D) Little diversity in artifact assemblage at the alleged quarry site, with many chunks and slabs exhibiting only cursory knapping features; (E) Large humped- or turtle-backed biface cores in various stages of production, with associated debitage of variable sizes (F) and possibly scant formal tools and micro debitage from occasional loss, discard, or resharpening of toolkits carried on-person into the quarry* (Stewart 1987:50).

Overall, the total assemblage appears to meet all Stewart's criteria for an ancient quarry, albeit a very small one.

Fire Evidence

A small shallow fire feature was unearthed (Fig. 8.40) at the top of excavation level 6, which was neither large nor deep. In fact, evidence of human habitation—hearths, ash pits, food remains, trash middens, or postholes—is lacking. There was also no evidence for natural vegetation sufficient for natural brush fires to occur in this partial enclosure. Therefore, the burned soil feature was likely produced from fire contact, not sustained burning. This evidence is consistent with the notion to assist quarrying through thermal spalling or heat-treating of the stone to promote easier fracture. More direct evidence supporting this claim remains on the east/northeast rock wall where an oval soot smudge remains visible (Fig. 8.41).

Attempts to Detach Rock with Large Chisels

On the same east/northeast wall, beginning just 10 cm beneath the excavated surface, eleven diagonal gouges running from top left to bottom right (at slightly arced slopes of about −1) began to be unearthed (Fig. 8.42). The longest gouge is about 62 cm in length while the shortest is only about 12 cm long and was unearthed at the lowest depth in level 6—below 76 cm of soil. The top half of this wall has linear depressions nearly horizontal (angled about 88° from vertical, 90° being truly horizontal). The linear depressions below the artificially cut gouges are slightly curved, but also run nearly 90° horizontal. The cut gouges themselves are mostly angled diagonally between 60 and 62° from 0° vertical. The cross sections of the five

Fig. 8.40 Excavation pit photo of Level 6 Feature 1: fire feature. This overhead view of the excavation pit depicts the shallow ~14 cm × 15 cm rectangular burned soil feature, which suggests evidence of localized fire use, perhaps for heat-treating lithics. The back wall of the excavation unit is visible at the bottom of the photo

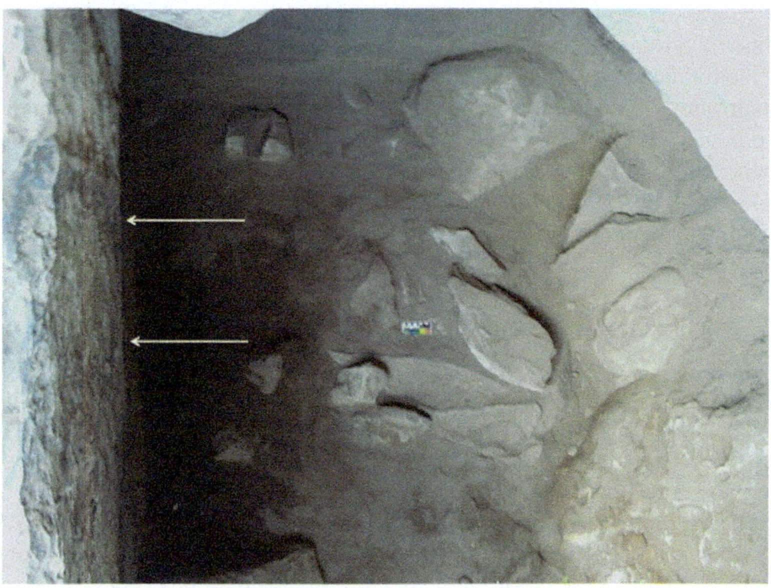

Fig. 8.41 Soot ring on east/northeast (quarried) wall where fire was probably used to weaken or spall pieces off the surface. Excavation pit floor shows the top of level 6

Fig. 8.42 Gigapan panorama of *Painel do Pilão* northeast excavation wall, which is part of the platform from which one can observe the rock paintings above. Gouges cut into it were buried to a depth of 70 cm from the original surface (top of excavation level 6)

northernmost (left side) gouges taper from millimeters wide at the top to about 4 cm wide at the bottom, and the cross sections of the remaining 5 gouges remain relatively parallel at about 3 cm width, with some variance between 2 to nearly 6 cm wide. The cross sections also vary in depth from a few millimeters to several centimeters, but all appear to have been cut into the wall directly perpendicular to the surface (at ~90° angles). The wall also displays shallow linear depressions that are somewhat horizontal. These shallow depressions suggest this was the general direction in which the wall was flattened, whereas the 60° gouges crosscut (overlay) those horizontal depressions. The gouges were intentional, after the wall was flattened.

Note that most gouges start narrow and widen to a rectangular shape at the bottom (Fig. 8.43 top). The five southernmost gouges to the right side abruptly terminate at a hard rocky protrusion sticking nearly 40 cm out from the wall. This rectangular ~30 × ~42 × ~2–6 cm thick protrusion itself is angled with an approximate slope of +1, nearly perpendicular to the gouges. Perhaps the stone cutters were unsuccessful at dislodging this protrusion, but any attempts to do so are not readily apparent. The central gouge begins just a short distance from where it ends, and it appears to be entirely rectangular at the edges (Fig. 8.43 bottom). This same feature supports the claim that some lithic tools unearthed in the excavation were used to cut stone.

Fig. 8.43 Close-up of the gouges cut into the northeast (platform/stage) wall of the excavation at *Painel do Pilão*; (top) depicts the initial propagation of some of the gouges. The termination (bottom) of some of the gouges that seem to have hit a hard protrusion

Because some gouges taper, it is unlikely that they were created by the natural erosion of softer rock or inclusions. Perhaps the longest gouges (at least) may have been made by swooping pendulum cuts with a massive stone. The fact that many of the tapered gouges run somewhat parallel to each other further supports the idea of several pendulum passes with a massive cutting stone, not the action of a single large boulder that naturally fell from a great height. A more massive stone would have required multiple people working in coordination to swing it or rig it so that it pivots around an axle—none of which were discovered here.

Alternatively, if the gouges were all chiseled or pecked from numerous repetitious actions, it was done with intense precision and effort. Supporting evidence for this is the fact that the *fourth gouge* begins and ends just within a short distance (Fig. 8.43 bottom). This gouge makes it more likely that it was chiseled, pecked, or gouged out instead of made by the motion of a pendular cutting tool. Closer inspection of the gouges also reveals slight cutting on the jutting protrusion underneath the gouges in the bottom image. The gouges slightly continue through the surface of the protrusion, further disproving that softer material naturally occurred there and eroded away long ago. The strongest theory currently is that a durable chiseling/gouging implement was used. Regardless, it seems significant planning and preparation were spent to create these cuts.

Damaged or Discarded Quarrying Tools Found in Related Context

Beginning at a depth of approximately 50 cm from the surface soil level (corresponding to excavation level 5), large conical flakes began to emerge that may have been used as picks, chisels, or drills, along with significantly larger boulders that appear to have been detached from the quarried wall. Also significant is the recovery of a chisel-like drill from this level (Fig. 8.44), which offers the strongest evidence of a quarrying tool found. In addition to the adze-like tools found near clustered flakes—mentioned in the previous section—level 7 contained another drill-like tool, or perhaps a pick or chisel (Fig. 8.45). Level 5 is where most of the diagonal gouges terminate, but the last gouge terminates in level 6, suggesting that the beginning of gouge production (and probably their completion) occurred within the level 6, 7, and 8 depositions—before the Younger Dryas. Radiocarbon dates for level 5 are 6000 years later, so it is unlikely that one gouge was made and then the

Fig. 8.44 Level 5—object 18: Chisel-like drill flake. The dorsal side (**a**), lateral side (**b**), and ventral side (**c**). The proximal end (top of all three photos) appears to be slightly deformed due to compressive impact consistent with hammering

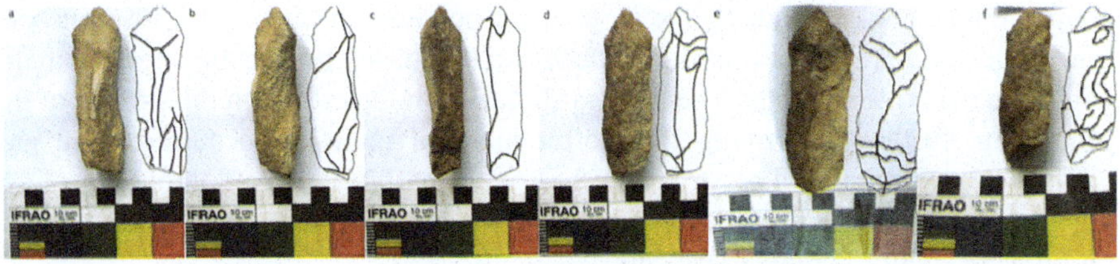

Fig. 8.45 Chisel-like drill from lithics in level 7 between boulders X and Y. The dorsal side (**a**), lateral right side (**b**), lateral left side (**c**), ventral side (**d**), oblique angle (**e**) showing distal end, and oblique angle (**f**) showing proximal end with a triangular cross section

others were made thousands of years later. Tools deposited at the bottom of level 5 might therefore have been dumped just before site abandonment, and the gouges, therefore, have the potential to be a clue correlated to the departure of the site.

Little Diversity in Artifact Assemblage/Cursory Knapping of Slabs

Many recovered lithics lacked blatant formal-tool features, but after washing and observing through low-level magnification, several cursory knapping techniques or uses were apparent. Although none of the recovered lithics match known contemporary styles (such as Clovis, Fish Tail, or Fell's Cave projectiles), there is evidence of direct and indirect hammer techniques applied to raw and heat-treated stone. Additionally, pecking, grinding, and in some cases, polishing were observed on many artifacts recovered from levels 5 through 8.

The surface treatments are unusual, as many lithics exhibit cursory characteristics. Interestingly, some treatments, like polishing seem to have concealed more obvious signs of cursory knapping, such as flake scars and retouch marks. This could be attributed to the varying quality of the stone, which ranged from coarse sand to cryptocrystalline consistency. Notably, some of the lowest-grade sandstones could even be altered simply by rubbing with one's thumb. However, evidence of on-site knapping attempts is represented by the discovery of a multifaceted core (Fig. 8.46) from level 6.

Large Humped- or Turtle-Backed Bifaces at Various Production Stages

Some of the wedge-shaped flakes from level 8 can be generally described as "humped." In addition to those that were discussed to have surface treatments, several also have a noticeable single flake removed from the dorsal planar surface—a flute (revisit Figs. 8.19, 8.26, and 8.27). Some also have faint imprints and indentations suggesting they were meant for hafting as well (more evident before the lithics were washed), and so the flute was possibly intended to enhance hafting efficiency of the lashings to the flattened surface (revisit Fig. 8.19 again). Unlike Clovis points, however, the flute is located near the middle of the dorsal surface, giving an overall appearance that can be described as "turtle-backed." Some turtle-backed flakes have no mid-dorsal flute (Fig. 8.47), but the surface displays a good deal of polish, nonetheless.

Fig. 8.46 Multifaceted core from level 6 object 11 is evidence that some knapping activity may have occurred on-site at the quarry. The left image shows several flake scars. The image on the right shows the flat underside. The two colorations formed naturally

Fig. 8.47 Turtle-backed large flake from level 5—object 6. The dorsal side (**a**) shows evidence of edge retouch just about all the way around, with flaking along most of the dorsal face. The flaking (or chipping) may have occurred from use as a scraper. The ventral side (**b**) had several short dark lines (before washing), possibly due to contact use or decayed lashings. Under 20× magnification, the "back" end on the dorsal side (**c**) appears to display an intentionally polished surface, compared to the edge (**d**) which appears to have use-wear damage (scraping?). Similarly, there also appears to be polish on the ventral surface (**e**), possibly acquired from scraping use

Scant Formal Tools, Micro Debitage, and Occasional Loss of Formal Tools

Significantly less quarrying evidence was recovered from strata above excavation level five (levels 1–4), supporting radiocarbon evidence of sporadic visitation to the site after the Paleoindian period. The few lithics unearthed in higher levels resemble discarded broken tools devoid of knapping debitage. Therefore, the higher levels indicate a shift in the way the space was conceived and utilized over time. Levels six through nine (55 cm total depth) have the most significant evidence of quarrying activity.

Although the few charcoal pieces that have been recovered from higher levels span several thousand years, the narrow date range for the lowest levels, coupled with the wall gouges reaching this depth, offers strong evidence that the majority of quarry activity and deposition occurred in a span of ~500 years, beginning ~13,000 years ago, making *Painel do Pilão* an early Paleoindian quarry, and the wall is possibly the earliest intentionally terraformed structure in the Americas.

Overall, the lower four levels: 6, 7, 8, and 9 contained the most numerous lithic artifacts recovered compared to the highest five levels (63% to 37%), the heaviest lithics recovered by weight (66% to 34%), and the largest volume of lithics recovered (22,600 cm^3 to 11,600 cm^3) despite excluding the massive boulders that were too large to be removed and weighed (see Table 8.3). Lithics predominate the lowest four levels despite being recovered from less overall excavated soil volume—1.58 m^3 compared to 2.36 m^3 in the highest levels (40% to 60% soil volume). Although many artifacts in the lowest levels have inconclusive functions, several of them have evidence of use (Fig. 8.48), or cursory knapping for intended use (Fig. 8.49).

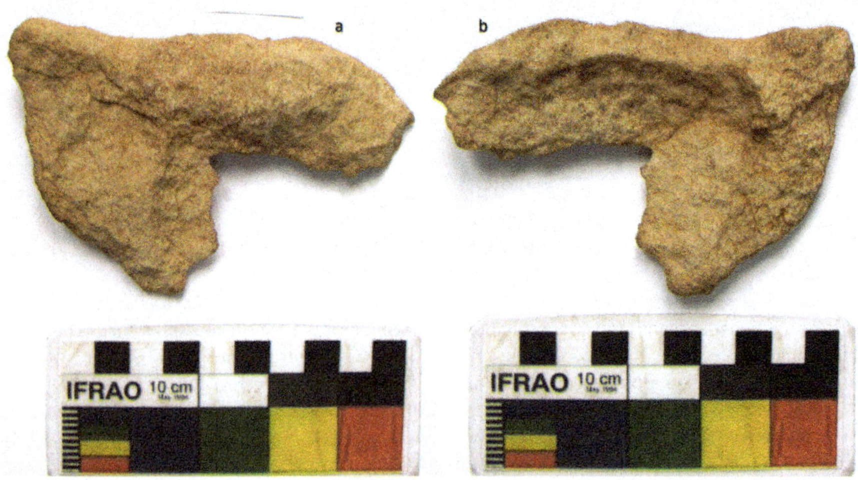

Fig. 8.48 Point-angled flake (object 15) from level 7. The dorsal side (**a**) and the ventral side (**b**) share an interior acute angle that is tapered, perhaps where it may have been affixed into a splint of wood. However, the intended function is unknown

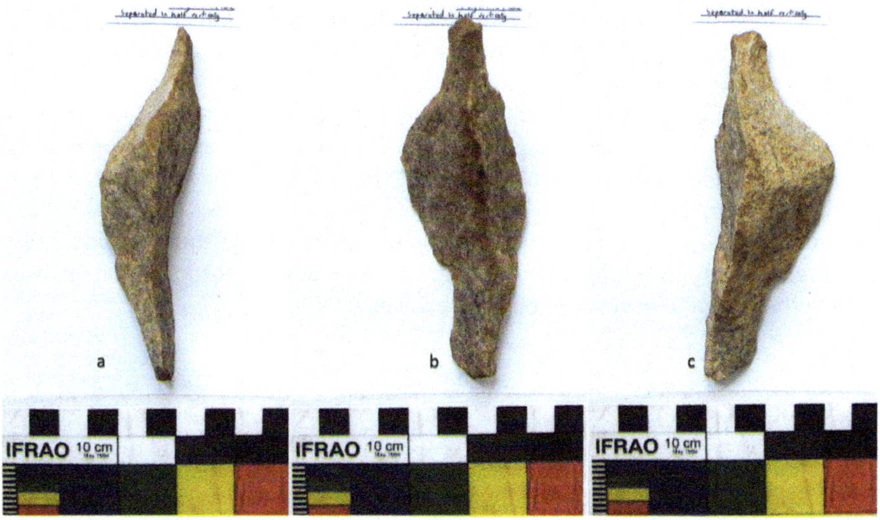

Fig. 8.49 Long, tetrahedral lithic from level 7—object 6. The lateral side (**a**) shows a proximal end tapered to a darkened triangle, and a broken distal tip. The ventral side (**b**) has a transverse midline ridge, and the dorsal side (**c**) has a whitish-to-beige coating that appears slightly polished

Assuming all the lithic litter in the excavation pit came from the quarry wall, and in estimating that the entire surface area of the wall was reduced, then on average at least 0.7 cm of the entire 180 cm × 260 cm wall was flattened (see Table 8.3). This estimation is the minimum amount the data provides. A seemingly large number of scooped-out wall flakes, and useful blanks, may have been carried away from the site in antiquity, as proposed previously. Table 8.3 includes all lithic materials recovered from excavation and screening, but it does not include data on boulders in the excavation pit that were too large to be removed and weighed. Utilizing the dates and soil depths of the total excavation pit, the calculated average sedimentation rate computes to ~4 cm soil depth per century.

Two possible scenarios, not mutually exclusive, could fit the overall lithic assemblage revealed from this excavation:

1. Many desired flakes were minimally prepared here for inspection of internal flaws, with satisfactory specimens transported to a final knapping worksite while those that failed were discarded here, along with worksite tools, some of which were painted. Utilized lithics were the probable tools used here to obtain new specimens, and/or later toward artistic/architectural purposes.
2. The intent was never for lithic sourcing but instead to reshape the rock shelter itself, but just the east/northeast wall. The tools were cursory, and paint used on the wall was similarly used on tools and flakes meant for the associated tasks. It is important to note, though, that no other part of the massive boulder outside the shelter appears to have been flattened, so it is unusual for so much work to be done on a side that would not have been viewed (or even noticed) while standing atop it.

Regardless of which scenario is correct, the fact is that human activity here is profusely infused with artistry (not sole ecological efficiency) as early as 13,000 years ago—a precedent for New World initial inhabitants. Paleoindian artisans here were less concerned with consistent lithic manufacture and utilitarian tool durability that scholars might argue is the case with better knapping materials like chert and flint. Instead, the culture here appears to have favored chromatic embellishments and monumental art. Why would artisanship eclipse utilitarian hunting preparedness in the tropical rainforest of the Amazon? We can surmise that a peaceful surge of daily leisure is a prerequisite for this transition—the inhabitants were not overburdened with survival pursuits each day.

Conclusion

Regarding its lithics, rock art, and terraformed wall, Painel do Pilão displays specific artistic adaptations distinct from traditional Paleoindian trends currently studied, but social factors at many locations have always been implied. Many of the lithics may have been used for wood and/or stone smoothing/abrading. Few were shaped as finalized projectile points or other functions related to food procurement or processing. Because virtually none of the 1688 excavated lithics show evidence contextually related to food procurement or processing, this site provides a new window into the lifestyles of Paleoindians beyond hunting and subsistence contexts—and it will therefore broaden our understanding of Paleoindian activities adapted to new regions at the end of the Ice Age.

Many scholars still assume that the further back in time a culture existed, the more tethered to utilitarian needs the society must have been. However, we know that several millennia later in the Americas, trade and acquisition of good quality, unique, or prestigious minerals was paramount in facilitating cross-continental interaction and ideological influence (Baugh and Ericson 2013). The Hopewell interaction sphere (Byers 2011; Charles and Buikstra 2002) of the Eastern Woodlands North America period is a prime example. Although trade requires maintained communication among groups, the motivations driving trade have probably remained consistent since we have evolved into anatomically modern humans.

Beyond the superior quality of many minerals to cleave predictably, some also intrigue us more than others, as they probably did to the ancients as well, due to their stark colors and exotic luster. However, Paleoindian studies almost exclusively focus on the mechanics of those minerals, and less on the aesthetic properties that provoke social aspirations.

The trend at *Painel do Pilão* speaks to the latter. The stratigraphic sequence of the excavation to bedrock reveals a potential early affinity toward naturally colored rocks, and in level 8 there are also highly polished unifacial flakes and boulders, some with intentionally painted surfaces. Based on artifacts from levels 5 to 7, preference for color seems to have transitioned into a painting tradition of their tools' polished surfaces—as well as on walls in the form of rock art that demarcated horizon events. All these clues are dated provocatively near the onset of the Younger Dryas cold period, which seems peculiar among reports of Paleoindians. In light of evidence provided at this site, scholars should take added consideration of social opportunities and rewards when reporting on Paleoindians alongside technological, environmental, or subsistence arguments (Fagundes et al. 2008; Newby et al. 2005).

How did Paleoindians perceive their time and efforts; were they preoccupied with survival and/or food procurement or are these painted lithics and the straightened wall clues suggesting that even as far back as 11 to 13 thousand years ago, people took a good deal of effort toward appeasing themselves with insubstantial, but personal, activities. Were they aware of (and preoccupied with) the environmental changes that were occurring? Was art considered a magical way to affect the inevitable changes in the ecosystem (ritual contagious and imitative magic), or was it simply to distract from reality? Were these Paleoindians simply trying to beautify their tools and shelters, or did they think these activities were magically effective?

Whenever a community is at ease with pursuits of subsistence, material resources, and safety, individuals can devote themselves to personal and social ambitions. Living in the caves of Monte Alegre's forested hills during the Terminal Pleistocene allowed for plenty of food, fresh water, stone, wood, and shelter during a time when chance encounters with other human groups were presumably low—especially if they did indeed establish a seasonal almanac through their rock art at the nearby sites of *Serra do Sol* and *Serra da Lua*. The amount of effort this first community divested here to terraform a wall and paint lithics implies that they were not inexorably preoccupied with just subsistence acquisition, but instead could transition to artistic, symbolic, and possibly even educational (landscape learning) activities.

References

Baugh, T. G., and J. E. Ericson. 2013. *Prehistoric exchange systems in North America*. Springer.

Byers, A. M. 2011. *Sacred games, death, and renewal in the ancient eastern Woodlands: The Ohio Hopewell system of cult sodality heterarchies*. Rowman Altamira.

Caletti, C. C. 2020. Göbekli Tepe and the sites around the Urfa Plain (SE Turkey): Recent discoveries and new interpretations. *Asia Anteriore Antica. Journal of Ancient Near Eastern Cultures* 2:95–123.

Charles, D. K., and J. E. Buikstra. 2002. Siting, sighting, and citing the dead. *Archeological Papers of the American Anthropological Association* 11 (1): 13–25.

Cheng, H., H. Zhang, C. Spötl, J. Baker, A. Sinha, H. Li, M. Bartolomé, A. Moreno, G. Kathayat, and J. Zhao. 2020. Timing and structure of the younger Dryas event and its underlying climate dynamics. *Proceedings of the National Academy of Sciences* 117 (38): 23408–23417.

Davis, C. S. 2009. Notícia De Pesquisa Em Andamento [Current research: Archaeoastronomy at Monte Alegre, Pará – A research problem and research strategy]. *Amazônica* 1 (2): 536–547.

Davis, C. S. 2014. *Archaeoastronomy of terminal Pleistocene rock art on the Amazon River at Monte Alegre, Pará, Brazil*. Chicago, IL: Doctoral Dissertation, Department of Anthropology, University of Illinois at Chicago.

Davis, C. S., A. C. Roosevelt, W. Barnett, and J. P. Brown. 2017. Paleoindian solar and stellar pictographic trail in the Monte Alegre Hills of Brazil: Implications for pioneering new landscapes. *Journal of Anthropology and Archaeology* 5 (2): 1–17.

De Lorenzis, A., and V. Orofino. 2014. New possible astronomic alignments at the megalithic Site of Göbekli Tepe, Turkey. *Archaeological Discovery* 3 (01): 40.

Fagundes, N. J. R., R. Kanitz, R. Eckert, A. C. S. Valls, M. R. Bogo, F. M. Salzano, D. G. Smith, W. A. Silva Jr., M. A. Zago, A. K. Ribeiro-dos-Santos, S. E. B. Santos, M. L. Petzl-Erler, and S. L. Bonatto. 2008. Mitochondrial population genomics supports a single pre-clovis origin with a coastal route for the peopling of the Americas. *The American Journal of Human Genetics* 82:583–592.

Funk, R. E., T. P. Weinman, and P. L. Weinman. 1969. The Kings Road Site: A recently discovered Paleo-Indian manifestation in Greene County, New York. *The Bulletin* 45:1–23.

Grier, C., B. Angelbeck, and E. McLay. 2017. Terraforming and monumentality as long-term social practice in the Salish Sea Region of the Northwest Coast of North America. *Hunter Gatherer Research* 3 (1): 107–132.

Haynes, G. 2002. *The early settlement of North America: The Clovis era.* Cambridge University Press.

Holliday, V. T. 2000. The evolution of Paleoindian geochronology and typology on the Great Plains. *Geoarchaeology* 15 (3): 227–290.

Newby, P., J. Bradley, A. Spiess, B. Shuman, and P. Leduc. 2005. A Paleoindian response to Younger Dryas climate change. *Quaternary Science Reviews* 24 (1): 141–154.

Randall, A., and K. Sassaman. 2017. Terraforming the middle ground in ancient Florida. *Hunter Gatherer Research* 3 (1): 9–29.

Rasmussen, S. O., K. K. Andersen, A. M. Svensson, J. P. Steffensen, B. M. Vinther, H. B. Clausen, M. L. Siggaard-Andersen, S. J. Johnsen, L. B. Larsen, D. Dahl-Jensen, M. Bigler, R. Roëthlisberger, H. Fischer, K. Goto-Azuma, M. E. Hansson, and U. Ruth. 2006. A new Greenland Ice Core chronology for the last glacial termination. *Journal of Geophysical Research* 111 (D06102): 1–16.

Reinig, F., L. Wacker, O. Jöris, C. Oppenheimer, G. Guidobaldi, D. Nievergelt, F. Adolphi, P. Cherubini, S. Engels, and J. Esper. 2021. Precise date for the Laacher see eruption synchronizes the younger Dryas. *Nature* 595 (7865): 66–69.

Roosevelt, A. C., M. Lima da Costa, C. L. Machado, M. Michab, N. Mercier, H. Valladas, J. Feathers, W. Barnett, M. Imazio, A. Henderson, J. Silva, B. Chernoff, D. S. Reese, J. A. Holman, N. Toth, and K. Schick. 1996. Paleoindian cave dwellers in the Amazon: The peopling of the Americas. *Science* 272 (5260): 373–384.

Stewart, R. M. 1987. Rhyolite Quarry and Quarry-related sites in Maryland and Pennsylvania. *Archaeology of Eastern North America* 15:47–57.

Tobolczyk, M. 2016. The World's Oldest Temples in Göbekli Tepe and Nevali Çori, Turkey in the Light of Studies in Ontogenesis of Architecture. *Procedia Engineering* 161:1398–1404.

Chapter 9
Solar-Aligned Pictographs at Painel do Pilao

Introduction

In the previous chapter, evidence was presented indicating that the northeast shelter wall had been intentionally knapped into a straight surface during the earliest period of human activity. The rubble, concentrated in the lowest excavation levels, dates narrowly to a ~500-year Paleoindian period, terminating at the onset of the *Younger Dryas*. The presence of such a flat (though not smooth) wall suggests that the rubble was not the result of natural wall collapse. If the unearthed rocks were simply flakes eroded from the wall, we would expect to find consistently thin or small lithic fragments with no signs of retouch or use.

Instead, the evidence highlighted in the previous chapter included variously coated, painted, and polished stone tools bearing use wear consistent with scraping, ablating, chiseling, and exposure to fire. Red, yellow, and white ochre, along with goethite and other pigment-bearing rocks, were also found in these lowest levels, pointing to their deliberate use as coloring agents.

Moreover, the significant number of utilized flakes bearing traces of paint makes it unlikely that ancient stone masons were merely removing old rock art. If this were the case, curvilinear fragments of stick figures, red and yellow dots, or other pictorial elements fractured across multiple rocks—like pieces of a puzzle—would have been discovered. Instead, the lithics and the flattened wall appear to have been intentionally crafted.

Previous theories offer partial explanations but fail to account for all the evidence. One theory posits that the shelter functioned as a sandstone quarry, with the most valuable stones removed from the site. Another suggests that the flattened wall was created not for its own utility but as a platform or stage to view the rock art above it, which remains visible today. Neither theory fully explains the extraordinary effort invested in flattening the wall itself. This chapter turns its focus to the

C. S. Davis, *Before Sunset*, Conflict, Environment, and Social Complexity,
https://doi.org/10.1007/978-3-031-93373-8_9

rock art positioned above the wall and introduces a new theory—one that explores the wall's potential role in relation to sunlight and shadows.

The wall from which rocks were removed also serves as a platform for observers to stand atop and view the red and yellow pictographs. Some of these paintings likely depict sun- and sky-themed representations due to their apparent solstice alignments. Among the most enigmatic pictographs is a grid pattern, with many of its boxes filled with "x" marks. As noted in the previous chapter, the pictographs follow a somewhat linear arrangement along a lower panel, while another sequence is positioned on an upper panel that angles skyward toward the northwest.

The outcrop is capped by tabular-to-platy sandstone, with an ~18° slope from horizontal, with a thin surface growth of greenish lichen. This sandstone cap gently protrudes over the rock art in an arc, shielding most of the paintings from precipitation and direct sunlight for part of the year. The combination of the solar alignment of these paintings, the dated antiquity of contextual artifacts, and the features uncovered during excavation suggests a long-standing tradition of naked-eye astronomy and pictographic rock art. The painted grid, in particular, may have served as one of the earliest solar "calendars" in the Americas.

Photographs and Brunton compass readings of the pictographs were taken in 2007 and 2008, and additional photos and enhanced photogrammetric techniques were taken in 2010 using a handheld GPS unit and a Gigapan panoramic robot, which preserves visual and angular ratios as seen from the platform stage (panorama also available online at http://gigapan.com/gigapans/45523). In 2011, before and during the excavation season, twenty prominent pictographs and 280 points on the outcrop surface were laser distance measured from the platform stage, as well as inclination readings, with a Leica Disto-8 laser distance measure.

Previous compass readings in 2007–2008 had to be corrected to true north, which for Monte Alegre's magnetic declination in 2008 was a correction of + 18°01'±19' [verified using (NCEI 2016)]. All electronic devices (GPS and distance measure) were set to automatically correct for true north at the time of reading; however, a variance of 0–5° intermittently reoccurred among the different devices. This variance may have been due to a combination of the following problems: human error, inclination of angle readings, and local electronic or magnetic interference—presumably from higher iron in the local (platform/stage) rock. Alerting the journal to a mix-up of some of the compass readings with true-north electronic readings was the reason for retraction of a previous article published in 2016. For this reason, compass readings were checked again during repeated trips in December of 2016, and August of 2023, including some independent verification.

All photographs were examined and categorized into motifs, like those done at *Serra da Lua*. A count of 230 pictographs were tallied with assistance using Dstretch image enhancements to distinguish overlapping drawings when possible. A table of the pictograph is provided in Table 9.1, sorted by location on the wall (southeast wall, for example). However, the total number of pictographs must be taken as a general guide. Much of the art here appeared more faded than at *Serra da Lua*, and the true intent and extent of some paintings might never be reconstituted.

Table 9.1 *Painel do Pilão* rock art statistics

ID	Location	Color	Count	Single or Panel	Motif	Surface
68	East Wall	Red	1	Panel, calendar	Grid	Flat
69	East Wall	Red	1	Panel, grid with faded outline?	Grid	Convex
70	East Wall	Red	1	Panel, grid with faded outline?	Meander	Convex
71	East Wall	Yellow	3	Panel, grid with faded outline?	Abstract	Convex
72	East Wall	Red	1	Panel, calendar	Abstract (zoomorph?)	Flat
73	East Wall	Yellow	1	Panel, calendar	Zoomorph (linear torso)	Flat
74	East Wall	Yellow	1	Panel, calendar	Grid	Flat
75	East Wall	Red	1	Single	Abstract (geometric)	Uneven
76	East Wall	Red	1	Below calendar	Face	Flat
77	East Wall	Red	1	Below calendar	Abstract (bug)	Flat
78	East Wall	Red	1	Below calendar	Face	Flat
79	East Wall	Red	1	Below calendar	Geometric (polygons)	Flat
80	East Wall	Red	1	Below calendar	Geometric (polygon)	Convex underside edge
81	East Wall	Red / yellow	3	Below calendar	Smudge	Flat
82	East Wall	Red	1	Below calendar	Geometric (trident)	Uneven
83	East Wall	Red	1	Below calendar	Circle	Convex
84	East Wall	Red	1	Below calendar	Geometric (crisscross)	Convex
85	East Wall	Red / yellow	1	Below calendar	Concentric dot	Slightly convex
86	East Wall	Yellow	1	Below calendar	Geometric (linear)	Slightly convex
87	East Wall	Yellow	1	Below calendar	Circle or dot rayed	Slightly convex
88	East Wall	Yellow	1	Below calendar	Smudge	Concave
89	East Wall	Red on yellow	1	Below calendar	Landscape?	Convex
90	East Wall	Red / yellow	1	Below calendar	Landscape?	Convex
91	East Wall	Red on yellow	1	Single, missing section	Abstract (left curly braces)	Concave
92	East Wall	Red on yellow	1	Single, missing section	Dot	Convex

(continued)

Table 9.1 (continued)

ID	Location	Color	Count	Single or Panel	Motif	Surface
93	East Wall	Red on yellow	1	Single, missing section	Geometric (backslash \)	Convex
94	East Wall	Red on yellow	1	Single, missing section	Arc	Convex
95	East Wall	Red on yellow	1	Below calendar	Landscape?	Convex
96	East Wall	Red / yellow	1	Below calendar	Zoomorph (linear torso)	Convex
97	East Wall	Red	1	Below calendar	Geometric (T)	Flat
98	East Wall	Reddish-orange	1	Right of calendar	Grid (lattice)	Flat
99	East Wall	Reddish-orange	1	Right of calendar	Face	Flat
100	East Wall	Red / yellow	1	Right of calendar	Face	Flat
101	East Wall	Red	1	Right of calendar	Geometric (notational?)	Flat
102	East Wall	Red	1	Right of calendar	Geometric (notational?)	Flat
103	East Wall	Red	1	Right of calendar	Zoomorph (oval torso)	Flat
104	East Wall	Red	1	Right of calendar	Lollipop	Flat
105	East Wall	Red	1	Right of calendar	Lollipop	Flat
106	East Wall	Red	1	Right of calendar	Zoomorph (dumbell torso)	Flat
107	East Wall``	N/A/ pink	1	White patch	Abstract	Flat
108	East Wall``	N/A/ red/ pink	1	White patch	Geometric (crisscross)	Convex around dark patch
3	Far South shelter	Red	1	Single	Circle	Convex ledge
4	Far South shelter	Red	1	Single	Geometric (v)	Flat, angled upward, floor
5	Far South shelter	Red	1	Single	Crescent	Flat, angled upward, floor
6	Far South shelter	Red	1	Single	Dot	Flat, angled upward, floor deep in shelter
7	Far South shelter	Red	1	Single	Abstract (geometric)	Convex curve below shelter
8	Far South shelter	Red / yellow	1	Panel, caminho do sol beginning	Circle rayed dispersed	Convex ledge, pockmarked alcove backdrop
1	Far South wall	Red	6	Panel of 6 or 7 vertical lines	Geometric (linear)	Flat

(continued)

Table 9.1 (continued)

ID	Location	Color	Count	Single or Panel	Motif	Surface
2	Far South wall	Red	1	Single	Dot	Concave alcove base
114	Floor	Red	1	Single	Abstract (geometric)	Flat
115	Floor	Red	1	Single	Abstract (geometric)	Flat
116	Floor	Red	1	Single	Smudge	Flat
117	Floor	Red (orangish)	1	Single	Smudge	Convex
118	Floor	Red	1	Single	Dot	Convex (corner)
119	Floor	Red	1	Single	Geometric (arrow head)	Flat
120	Floor	Red	1	Single	Geometric (arrow head)	Flat
121	Floor	Red	1	Single	Geometric (arrow head)	Flat
9	South wall	Red	2	Single	Circle / alcove	Alcove rim
10	South wall	Red	2	Panel, small group of alcoves	Circle / alcove	Alcove rim
11	South wall	Red	1	Single	Circle	Alcove rim
31	Southeast	Yellow	3	Panel, three insignias of same design, size, and wall height	Insignia	Flat
32	Southeast	Yellow	3	Panel, three insignias of same design, size, and wall height	Insignia	Flat
33	Southeast	Red	1	Single	Humanoid linear torso	Flat
34	Southeast	Red	2	Single	Zoomorph (bar torso)	Slightly concave
35	Southeast	Red	1	Single	Grid	Flat
36	Southeast	Red	1	Panel, linear row	Abstract (script?)	Convex
37	Southeast	Red	1	Single	Smudge	Convex
38	Southeast	Red	1	Single	Dot	Convex
39	Southeast	Red	1	Single	Smudge	Convex
40	Southeast	Red	4	Panel	Smudge	Convex
41	Southeast	Red	1	Single	Zoomorph (lobed torso)	Convex
42	Southeast	Red	1	Single	Geometric	Flat
43	Southeast	Yellow	5	Panel, five or maybe six handprints in a row	Handprints	Slightly convex

(continued)

Table 9.1 (continued)

ID	Location	Color	Count	Single or Panel	Motif	Surface
44	Southeast	Yellow	1	Single	Handprints	Slightly concave
45	Southeast	Yellow	4	Panel of four handprints in a row	Handprints	Flat
46	Southeast	Yellow	2	Panel of two handprints, a pair	Handprints	Slightly convex
47	Southeast	Red (yellow "static")	1	Single	Humanoid (bent shaft torso)	Convex
48	Southeast	Red (yellow "static")	8	Panel, matrix of dumbbells	Lollipop	Convex
49	Southeast	Red (yellow "static")	2	Panel, matrix of dumbbells	Lollipop	Convex
50	Southeast	Red	1	Panel, matrix of dumbbells	Fireball	Flat
51	Southeast	Red	1	Panel, matrix of dumbbells	Fireball	Flat
52	Southeast	Red	1	Panel, matrix of dumbbells	Geometric (linear)	Flat
53	Southeast	Red / yellow	1	Panel, landscape map?	Abstract (landscape map?)	Convex underside
54	Southeast	Red / yellow	1	Panel, landscape map?	Circle / geometric (line)	Convex underside
55	Southeast	Red / yellow	1	Panel, landscape map?	Grid	Convex underside
56	Southeast	Red	1	Single	Grid	Convex
57	Southeast	Red	1	Single	Abstract (grid?)	Convex underside
58	Southeast	Red	2	Panel, matrix of dumbbells	Lollipop	Convex
59	Southeast	Red	2	Panel, matrix of dumbbells	Lollipop	Convex
60	Southeast	Reddish-orange	1	Panel, matrix of dumbbells	Zoomorph (oval torso)	Convex
61	Southeast	Red	1	Panel, matrix of dumbbells	Abstract (grid?)	Flat
62	Southeast	Reddish-orange	1	Panel, matrix of dumbbells	Scene	Convex
63	Southeast	Red	2	Panel of two handprints	Handprints	Flat, angled downward
64	Southeast	Red	1	Single	Geometric	Slightly convex
65	Southeast	Red	1	Single	Insignia	Convex underside edge

(continued)

Table 9.1 (continued)

ID	Location	Color	Count	Single or Panel	Motif	Surface
66	Southeast	Red	1	Single	Abstract (indeterminable)	Flat
67	Southeast	Red	1	Single	Handprints	Convex (corner)
12	Southeast	Red	1	Single	Geometric (linear)	Slightly convex
13	Southeast	Red	1	Single	Geometric (polygon)	Slightly convex
14	Southeast	Red	1	Single	Arc	Slightly convex
15	Southeast	Red	1	Single	Geometric (pipe)	Slightly convex
16	Southeast	Red	1	Panel, small abstract group	Abstract (bug)	Slightly convex
17	Southeast	Red	1	Panel, small abstract group	Abstract (script?)	Slightly convex
18	Southeast	Red / yellow	1	Single	Abstract (script?)	Bi-planar across vertex
19	Southeast	Red / yellow	1	Single	Abstract (script?)	Flat, angled downward
20	Southeast	Red	2	Panel, small two or three separate grids	Grid	Flat, angled downward
21	Southeast	Red	1	Panel, three or four overlapping designs	Abstract	Convex
22	Southeast	Red	1	Panel, three or four overlapping designs	Abstract	Concave shelf
23	Southeast	Red	1	Panel, three or four overlapping designs	Grid	Convex
24	Southeast	Yellow / orange	1	Panel, three or four overlapping designs	Abstract (script?)	Convex
25	Southeast	Red	1	Single	Fishhook	Slightly concave underside
26	Southeast	Red	1	Single	Grid	Uneven
27	Southeast	Red	1	Single	Grid	Uneven
28	Southeast	Red	1	Single	Grid	Uneven
29	Southeast	Red	1	Possible panel	Dot	Flat
30	Southeast	Red	1	Possible panel	Abstract	Convex
109	Underside	Red	1	Below calendar	Abstract	Concave alcove base
110	Underside	Red / yellow	2	Below calendar	Abstract (complex)	Concave alcove base
111	Underside	Red	1	Below calendar	Abstract (geometric)	Concave alcove base

(continued)

Table 9.1 (continued)

ID	Location	Color	Count	Single or Panel	Motif	Surface
112	Underside	Red	1	Below calendar	Abstract (geometric)	Concave alcove base
113	Underside	Red	1	Below calendar	Circle	Concave alcove base
157	Upper deck ceiling	Red	1	Single	Handprints	Convex
158	Upper deck ceiling	Red	1	Single	Abstract (geometric)	Uneven
159	Upper deck ceiling	Red	1	Single	Abstract (geometric)	Convex
160	Upper deck ceiling	Red	2	Single	Handprints	Convex
161	Upper deck ceiling	Red	1	Single	Smudge	Convex
162	Upper deck ceiling	Red	3	Single	Abstract (complex)	Convex
163	Upper deck ceiling	Red (orangish)	1	Single	Abstract (geometric)	Convex
164	Upper deck ceiling	Red (orangish)	1	Single	Humanoid (bent shaft torso)	Slightly convex
165	Upper deck ceiling	Red (orangish)	2	Single	Smudge	convex
166	Upper deck ceiling	Red	1	Single	Abstract (zoomorph?)	Convex
167	Upper deck ceiling	Red	2	Single	Abstract	Convex
168	Upper deck ceiling	Red	1	Single	Insignia	Convex
169	Upper deck ceiling	Red	1	Single	Geometric (wave)	Convex
170	Upper deck ceiling	Red	1	Single	Geometric (polygons)	Flat
138	Upper deck floor	Red / yellow / white	5	Single	Circle or dot rayed	Convex (corner)
139	Upper deck floor	Red	1	Single	Abstract (bug)	Convex ledge
140	Upper deck floor	Red / yellow / white	2	Single	Grid	Convex ledge/ step/corner
141	Upper deck floor	Red	2	Single	Smudge	Flat, angled upward, floor
142	Upper deck floor	Red	1	Single	Smudge	Flat, angled upward, floor
143	Upper deck floor	Red	2	Single	Spiral	Convex

(continued)

Table 9.1 (continued)

ID	Location	Color	Count	Single or Panel	Motif	Surface
144	Upper deck floor	Red	1	Single	Abstract (zoomorph?)	Convex (corner)
145	Upper deck floor	Red	1	Single	Abstract (indeterminable)	Convex
146	Upper deck floor	Red	1	Panel, asterisk and teacher panel	Arc	Convex edge
147	Upper deck floor	Red	1	Panel, asterisk and teacher panel	Asterisk	Uneven
148	Upper deck floor	Yellow	1	Single	Dot	Convex ledge/ step/corner
149	Upper deck floor	Red/yellow	4	Panel, asterisk and teacher panel	Abstract (complex)	Uneven
150	Upper deck floor	Red	1	Panel, asterisk and teacher panel	Asterisk / circle	Uneven
151	Upper deck floor	Red / yellow	1	Panel, asterisk and teacher panel	Zoomorph (oval torso)	Flat, angled upward, floor deep in shelter
152	Upper deck floor	Red	1	Panel, asterisk and teacher panel	Insignia	Flat, two planar, angled upward and outward, floor edge
153	Upper deck floor	Red	1	Panel, asterisk and teacher panel	Concentric circle	Flat
154	Upper deck floor	Red	1	Panel, asterisk and teacher panel	Insignia	Flat, angled upward, floor deep in shelter
155	Upper deck floor	Red / yellow	1	Panel, asterisk and teacher panel	Scene (the teacher?)	Uneven
156	Upper deck floor	Red / yellow (orange)	4	Single	Grid	Concave alcove base
125	Upper deck ledge	Red / yellow (orange)	1	Single	Abstract (complex)	Convex
126	Upper deck ledge	Red	2	Panel of two handprints, a pair?	Handprints	Convex
127	Upper deck ledge	Red	1	Single	Abstract (glasses)	Slightly concave
128	Upper deck ledge	Red / yellow (orange)	1	Single	Abstract (handprint)	Convex
129	Upper deck ledge	Red	1	Single	Lollipop	Convex

(continued)

Table 9.1 (continued)

ID	Location	Color	Count	Single or Panel	Motif	Surface
130	Upper deck ledge	Red	2	Panel, spider-finger handprint pair	Abstract (handprint)	Convex
131	Upper deck ledge	Red	1	Single	Abstract (polygon)	Convex
132	Upper deck ledge	Red	1	Single	Meander	Convex
133	Upper deck ledge	Red	1	Single	Abstract (polygon)	Convex
134	Upper deck ledge	Red	1	Single	Lollipop	Convex
135	Upper deck ledge	Red	1	Single	Lollipop	Convex
136	Upper deck ledge	Purple	1	Panel, of two larger images and smaller images	Abstract (zoomorph?)	Convex
137	Upper deck ledge	Purple	1	Panel, of two larger images and smaller images	Zoomorph (rectangle torso)	Convex
122	Upper Southeast Wall	Red	1	Panel, caminho do sol	Zoomorph (lobed torso)	Convex ledge, pockmarked alcove backdrop
123	Upper Southeast Wall	Red	1	Panel, caminho do sol	Zoomorph (lobed torso)	Convex ledge, pockmarked alcove backdrop
124	Upper Southeast Wall	Red	1	Panel, caminho do sol	Dot	Convex ledge, pockmarked alcove backdrop

The context of the excavation data gives the impression that the northeastern wall of the excavation unit was made flat in relation to it being part of a platform stage from which an observer can view the rock art on the outcrop. The platform offers a specific location for making observations at *Painel do Pilão*. When doing so, the painting most central to the observer's field of view is the grid image that has individual boxes marked with mostly repetitive (but some varying) tallies (Fig. 9.1). Roosevelt was the first scholar to consider its potential function as a calendar. Although alternative theories not related to astronomy cannot be ruled out, I explored several calendar theories (solar, lunar, planetary, stellar) and for years I focused on a possible, though not exact, pattern emerging with the intersection of the setting sun to a rocky gap in the near distance above and to the right of the painted outcrop (Fig. 9.2). The gap may have served as a "sun-capture" window, marking the afternoons when the sun intersects it. This chapter builds on that theory to add a few new possibilities of sun observations related to other art nearby, and the grooves on the excavation wall of the shelter relating to sun rays and shadows.

Fig. 9.1 Tally Grid "Calendar." The grid contains 49 visible contiguous boxes in 8 complete columns (plus a 9th incomplete one) of 5, 6, and 7 different-sized rows

Fig. 9.2 *Painel do Pilão* Gigapan panorama of the platform in the foreground (plus water bottle and camera bag), the grid ~9 m away, and the rocky gap, later laser measured 45 m away in the top right corner

Reading the "Grid"

The discernible grid yields 49 total squares, as shown in the first row of Table 9.4. Thirty of those boxes are filled with full X's, five of which have additional marks resembling asterisks instead of X's. Beyond these 30, four other boxes have ¾ of an X, five boxes are filled with a single vertical line, three boxes are filled with forward slashes, one box is filled with a backward slash, and four boxes appear to be empty.

The enhanced-color photo (particularly on the top right side of Fig. 9.3) makes visible that some horizontal grid lines appear to overlay physical lines, engraved or naturally existent, on the rock surface. At least the top row of the grid image seems to follow a linear feature in the wall, suggesting that the painter(s) of the grid utilized preexistent physical lines on the vertical surface. Beyond that last column-like line, the enhanced image shows that the grid transitions into more of a honeycomb lattice pattern. However, the last potentially incomplete column does provide an important clue that the grid was constructed sequentially from left to right, and therefore should be read in that same direction.

Another obstacle, though, is whether the ancient artist(s) intended to "read" the grid from left to right and top to bottom of each column? Or was it from bottom to top, resetting at the bottom for each column? Or did they read row by row instead of column by column? To figure this out, I tallied each possibility to determine which direction yielded the strongest pattern—the direction that yielded the most patterned sequence. Each type of pictographic "tally mark" was given a number

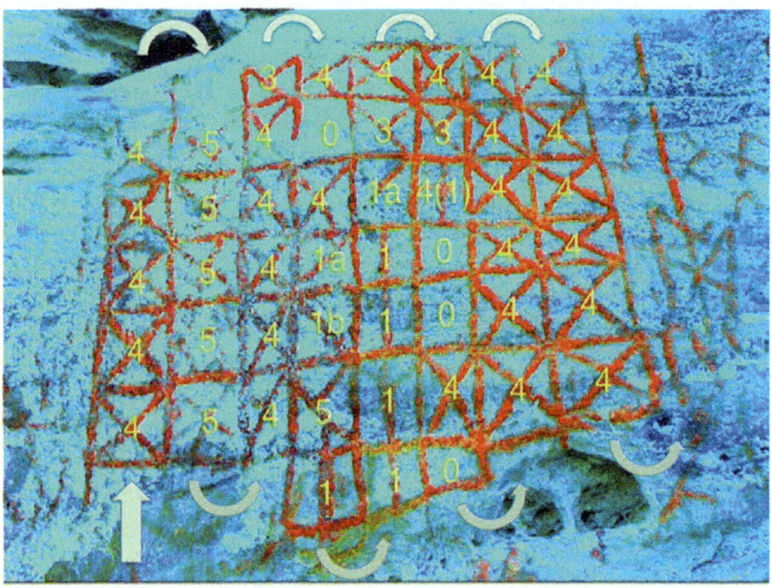

Fig. 9.3 *Painel do Pilão* grid image enhanced with Photoshop lab mode and assigned numbers for tally mark styles. Arrows indicate proposed method for "reading" the grid, based on notational clues and sequential style pattern recognition

from 0 to 5; then each possibility was recorded by sequence of numbers. Tables 9.2, 9.3, 9.4, 9.5 and 9.6 represent the sequence of grid marks recorded, with no need to include readings from right to left because that only reverses the order of the listed sequences. Any of the three configurations in the tables led to several assigned values separated by gaps that did not form reasonable patterns. Finally, the sinusoidal reading pattern (Table 9.6) demonstrated by the arrows in Fig. 9.3 was found to be the most meaningful pattern because nearly all the different tally marks are grouped together, particularly the tallies assigned with the number "1" (single vertical line) because they are in the interior of the grid, and reading them in any other way separates the 1s. Support of this hypothesis can be deciphered from the fact that the lower left corner of the grid, where I suggest one should start "reading" the sequence, has a yellow painting of a sinuous snake (visible in Fig. 9.4 with the black arrow pointing to it).

Table 9.2 Classification of "grid tally marks"

Clearly	Total boxes	Full X'S	3/4 X'S	Vert |	Forward /	Back \	Empty
Visible	49	30	4	5	3	1	4
Faded or Implied	7	4	1	1	1	0	5
Theoretical Total	54	34	5	6	4	1	9

Table 9.3 "Grid tally marks" by column

	Top->bottom; left->right	Cell gaps
≥ 5 Segmented	5,1	5,11, 27
4 Segmented	5,6,1,1,1,(1),1,12	0,6,1,4,6,1,(4),2,1,0
3 Segmented	1,1,1	10,13,6,17
Vertical segment	1,4,(1)	22,3,(2),19,(16)
Forward slash seg.	1,1	19,5,23
Backslash seg.	1,	20,28
Empty	1,2,1	17,15,1,12

Table 9.4 "Grid tally marks" by row

	Left->right; top->bottom	Cell gaps
≥ 5 Segmented	1,1,1,1,1,1	7,7,7,7,7,1
4 Segmented	6,1,3,2,4,1,3,1,3,1,3	1,1,3,1,1,1,3,1,3,1,2,3
3 Segmented	1,2,	0,9,37
Vertical segment	(1),1,1,1,2	(19),25,7,7,3,1
Forward slash seg.	1,1,	18,6,23
Backslash seg.	1,	33,15
Empty	1,1,1,1	9,17,7,12,0

Table 9.5 "Grid tally marks" by type

	Column count L->R	Row count T->B
≥ 5 Segmented	0,5,0,1,0,0,0,0	0,1,1,1,1,2,0
4 Segmented	5,0,5,2,1,3,6,6	5,4,6,4,4,5,0
3 Segmented	0,0,1,0,1,1,0,0	1,2,0,0,0,0,0
Vertical segment	0,0,0,1,4,0,0,0	0,0,0,1,1,1,2
Forward slash seg.	0,0,0,1,1,0,0,0	0,0,1,1,0,0,0,
Backslash seg.	0,0,0,1,0,0,0,0	0,0,0,0,1,0,0
Empty	0,0,0,1,0,3,0,0	0,1,0,1,1,0,1

Table 9.6 "Grid tally marks" by sinusoid

Sinusoidal sequence	4,4,4,4,4,5,5,5,5,5,4,4,4,4,4,3,4,0,4,1A,1B,5,1,1,1,1,1,1A,3,4,4,3,4,0,0, 4,0,4,4,4,4,4,4,4,4,4,4,
Starting at bottom left	4,4

Fig. 9.4 *Painel do Pilão* grid painting, Dstretch_ybr enhanced to enhance yellow pigment. The arrow points to a possible sinuous snake originally in yellow

Table 9.2 displays the total number of boxes at 49, the potential further number of boxes, and the different types of marks categorized as full X's, partial, or even empty.

Cell by cell sequence by reading columns. Table 9.3 sequences the different box-mark categories as they appear when "reading" the grid by column: from top to bottom, going from left to right.

Cell by cell sequence by reading rows. Table 9.4 sequences the different box-mark categories as they appear when "reading" the grid by row: from left to right, then top to bottom.

Tally mark count per grouping of "tally mark" types. Table 9.5 looks at the pattern that emerges (or doesn't) by seeking just one box-mark category at a time, counting how often that mark (re)appears moving column by column, and then moving row by row.

Table 9.6 sequences the different box-mark categories as they appear when "reading" the grid in a sinusoidal pattern from the bottom left column up, then going down the second column, then continuing up the third column, and so forth. This pattern is similar to the "baustrophedon" style of reading Mixtec codices in tenth-century Mesoamerica (Jansen 1991; Schaeffer 2017).

Archaeoastronomy Analysis of the Grid

The grid itself faces 16° northeast (corrected for true north). With their back to the grid, an ancient artist could observe the sunrise behind their right side and the sunset behind their left shoulder during the southern winter solstice. As the days progressed toward the northern summer solstice, the sun's apparent motion would shift more in front of the observer, with the sunset occurring in front of their left shoulder. I theorize that the circles connected to vertical lines, painted on both sides of the grid, provide additional evidence that the artist(s) were tracking the movement of the sun.

If the blank boxes, labeled as "0's," represented heavily overcast days when sky observations were indeterminable, the sinusoidal sequence supports the solstice hypothesis. If the grid's ancient artist(s) were observing the sun's passage as it set behind *Painel do Pilão* each day, they would have noticed the sun wandering slightly north or south on its journey to and from the southern winter solstice.

Full boxes, whether marked with asterisks or other symbols, might indicate days when the sun did not intersect with the rock window in the distance. Slash marks could represent near intersections, while vertical slashes might indicate full intersections. Though the rock window has likely eroded significantly over thousands of years, making it wider than it was originally, the sun's intersection with the remaining window is still testable.

Laser measurements of the "sun window" were taken from about 1.8 meters[1] off the ground of the platform in front of the grid painting. From there, the "sun window" is about 45 meters away, its center lies 251° west, its angular elevation extends from 37.9° to 40.3° altitude (39.25° dead center), and it has an angular separation of about 3° horizontally.

The grid contains 49 visible, contiguous boxes in 8 columns of 5, 6, and 7 rows. An additional fish-net lattice pattern continues on the right-hand side, suggesting that this column was the last one added. The true-color photo shows that the painting is in red pigment, but D-stretch image enhancement reveals that yellow pigment appears to overlap some red lines (revisit Fig. 9.4), especially in the four columns

[1] An arbitrary height at 1.8 meters (5'9").

on the right-hand side. A sinuous yellow image resembling the snake motif, closely resembling a red pictograph at Serra do Sol, appears in the lower left corner of the grid. In some places on the grid, red lines appear to overlay yellow, and in other places, yellow overlays red, making it difficult to conclude which color was original. Perhaps the dual use of color carries symbolic significance currently unknown.

D-stretch also enhances one or a few faded X-shaped tally marks in some of the central columns that currently appear as vertical lines. This suggests that the image has faded or was revised sometime in antiquity. Underlying the entire grid are remnants of a white backdrop. The preserved parts conspicuously coincide with most (but not all) of the boxes that have a full red "X" in them. As noted in the previous chapter, pigment nodules of these same three colors—red, yellow, and white—were unearthed in Level 7 (see Chap. 8: Evidence for Applied Pigment), which predates the Younger Dryas. Beneath the white backdrop are perhaps the remnants of 4 or 5 dark horizontal bands, which might be the natural coloration of the stone, either naturally banded or contrasted with bands of faded white color.

The intentional dual use of red and yellow is evident from a different pictograph at *Painel do Pilão* which is still visible to the naked eye (Fig. 9.5). This image appears to be an intentional composite of the two colors in continuous symbolism. Although categorized as an "abstract" image, the position of grids, triangles, and arcs in this pictograph might hold logographic information. The painting appears

Fig. 9.5 Red and yellow composite pictograph at Painel do Pilão photographed as is visible to the naked eye (top) and Dstretch_cb_lre enhanced (bottom) to highlight red and yellow details in color-balanced contrast

low on the southeast wall of the lower sequence of art. Depending on the direction in which the entire outcrop was approached in the past, this might have been among the first painted images that greeted those upon arrival. Due to the massive size of the platform relative to the "ground" today, it is from this direction that most visitors first view the rock art.

The position of arrival, departure, and observation is critically important when considering features or art that may have aligned to sky phenomena. Such sightings typically require some kind of fixed location from which to make observations. For the grid image at *Painel do Pilão*, the platform is an obvious location, and there is a lighter-colored gray rostrum that may have been meant to pinpoint the exact observation spot (Fig. 9.6). The annual movement of the sun relative to the rocky gap is visible from this spot, as is the grid, allowing for the intersection of the setting sun through the center of the gap to be relayed to the artist tallying the grid (Fig. 9.7). The gap itself might intentionally have been altered to specify this purpose. The perch is reachable by a short jaunt around a back trail (Fig. 9.7—top right). From the gray rostrum, the rocky perch runs in the direction 251° west, and at an inclination of ~39.25°. Obtaining these variables had remained challenging for years, especially due to correcting for true north, having a margin of error of a few cardinal degrees among electronic equipment, and due to difficulty measuring direct inclination and distance (Fig. 9.7—top left). Lacking the ability to camp on the outcrop for 49 days to observe by first-hand account leaves indirect measurements and calculations as the only means to verify the sun window theory.

Fig. 9.6 Potential observation spot of the grid image on the platform/stage at *Painel do Pilão*. The light gray "rostrum" where the author is standing may have been used as such. Author pictured in photo

Fig. 9.7 Potential sun-capture window at *Painel do Pilão*. If the pictographic grid is a tally of days in which the sun intersected the rocky gap in the distance, the top left image shows the direction of magnetic north in the compass, and the (green) correction line to true north. Top right image shows an assistant approaching the rocky gap from above

If each box of the grid represents one day of observing the setting of the sun relative to the rocky gap, then the artists recorded 49 days on or around the southern solstice (occurring on December 21 today). The sun intersects the window on its way to the solstice, then it comes back through the window as it moves towards the equinox (Fig. 9.8). Note how the tree branch serendipitously frames the sun as a "hitching post" on the day of the southern solstice (left photo). That tree did not exist 13,000 years ago, but perhaps something else was placed and used for a similar purpose. The grid painting has 49 total boxes, and the center boxes have tally marks that are simply vertical lines; most other tally marks are crisscrosses.

The fewest number of boxes are empty boxes with no tally marks inside. Perhaps these were cloudy days when no observations could be made. The discontinuation of the grid's right-hand column could be due to the obvious recognition of the sun moving further away from the gap. After 49 tallied days, the sun sets too far to the

Fig. 9.8 Above photos showing the sun intersecting the "rock window" (center) might have inspired ancient artists to tally the number of days between this event and the southern solstice sunset (left—taken December 21, 2008) to track the movement of the sun in a solar year (a calendar). The sun completely misses the rock window on the afternoon of the equinox (right—taken March 21, 2011)

Fig. 9.9 Faint image to the left of the grid is positioned where the tally of the sun would correspond to the solstice (left photo). The same photo enhanced with Dstretch_cb_lrd to highlight red and white colors shows more detail of what seems to be a face with an open, half-oval-shaped mouth

north, entirely missing the structure. If this is the correct interpretation of the grid, the yellow snake motif and the faint red face-like drawing are placed in the "solstice" position at the left of the grid (Fig. 9.9). The faint image has an appearance like a face with an open, toothed mouth. The mouth is conspicuously missing the white background, in the shape of a half oval. Perhaps the drawing symbolizes the sun dipping behind the profile of the outcrop on the day of the solstice, or perhaps on the day of the sun capture (compare to Fig. 9.8-middle). Could the rock window have looked like this faded image 13,000 years ago? A shallow, half oval is positioned at the bottom of the right column as well. Further beyond the grid is a red dot on a vertical line, possibly reflecting the sun as it entirely misses the rock window after day 49.

I previously theorized that the sun intersecting with the rocky gap may have been indicated by the vertical lines and that this would occur between 18–20 days before and after the solstice. However, SunEarthTools.com (2024) has equipped me with site-specific sun observations that I could not conduct in my limited visits to the site.

The data supports that the sun intersects the 39.25°-angle of inclination in the 251° cardinal direction of the rostrum on the platform/stage at 48 days before (Fig. 9.10) and 47 days after (Fig. 9.11) the southern (winter) solstice. These dates are graphically represented because they specifically coincide with the compass and laser distance recordings onsite at the upraised "rostrum" on the platform nearly 12 m away from the grid painting (revisit Fig. 9.6). My previously published observations were conducted on the platform approximately 9 m from the grid. The platform is large enough that the observer's position can introduce slight variations in measurements. However, given the brightness of the sun, it may be more accurate to describe the general range of dates during which the sun intersects the rock window. This range

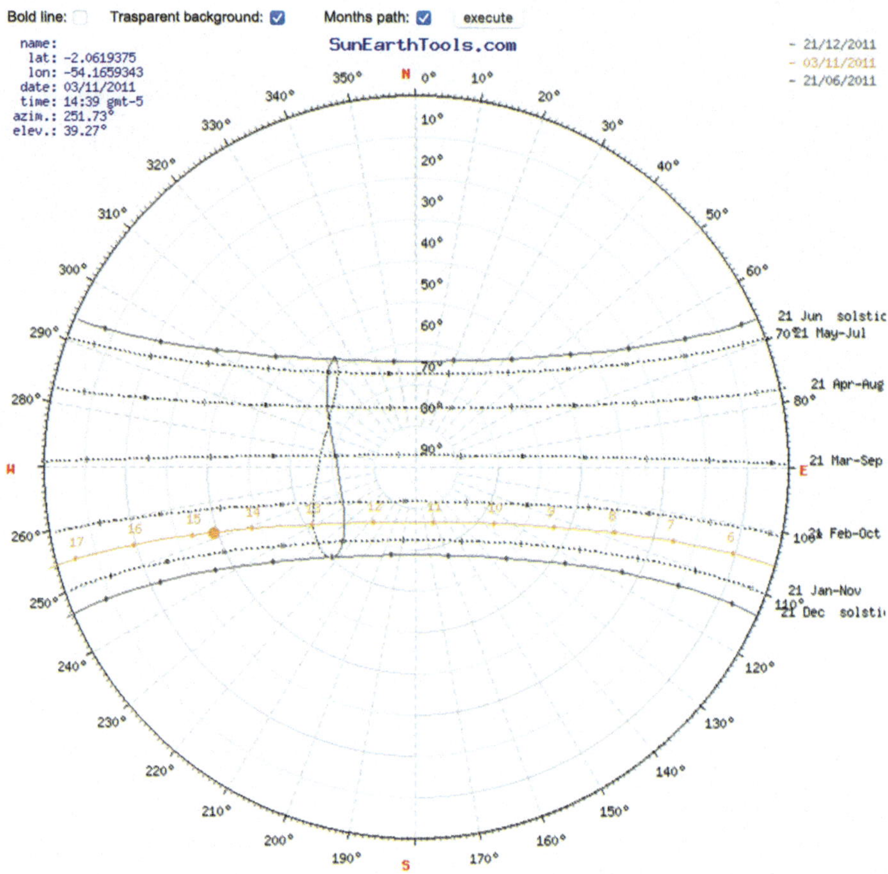

Fig. 9.10 Sun Chart of November 3, 2011, at *Painel do Pilão* (depicted as 03/11/2011 in the top left corner). The time is 2:39 PM (depicted as 14:39 gmt). The local azimuth (compass heading) of the sun is at 251.73° with an elevation (inclination) of 39.27°. The right top corner displays the solstice dates (21/12/2011 and 21/06/2011 correspond to Dec 21 and June 21). The yellow line indicates the path of the sun on the day of November 3 while the yellow dot is the position of the sun to the minute. The inner circles indicate the angle of the sun in the sky

appears to fall between ~48 days and ~18–20 days before and after the southern solstice. During this time frame, the sun passes through the rock window, potentially including days when it moves behind the side edges of the opening. It is important to note that the rock window has likely eroded over time, and the rostrum may not have been the precise observation spot used in the past.

A table has been included to illustrate how the sun's position in the sky changes on subsequent days. It should be noted that instrument measurements might contain slight inaccuracies due to mechanical variance and human error. For borderline days with imprecise data, values recorded a minute before or after the observation are provided in parentheses. These values highlight the minute changes in the sun's movements and the angles involved within such a brief span of time.

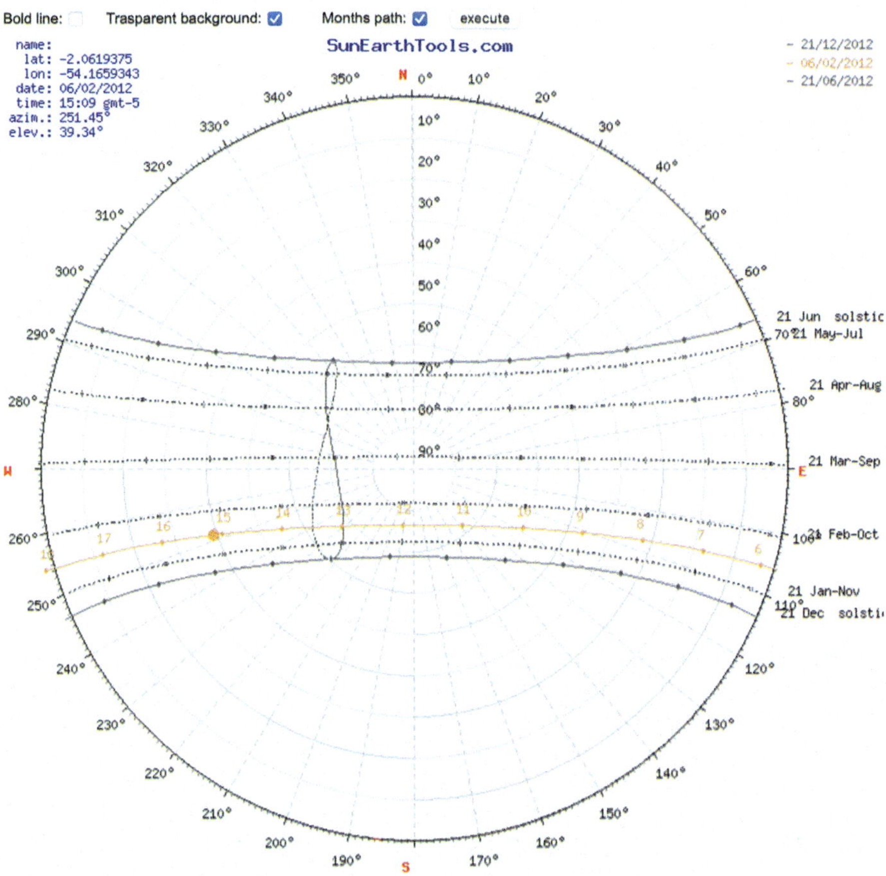

Fig. 9.11 Sun Chart of February 6, 2012, at *Painel do Pilão* (depicted as 02/06/2011 in the top left corner). The time is 3:09 PM (depicted as 15:09 gmt). The local azimuth (compass heading) of the sun is at 251.45° with an elevation (inclination) of 39.34°. The right top corner displays the solstice dates (21/12/2012 and 21/06/2012 correspond to Dec 21 and June 21). The yellow line indicates the path of the sun on the day of February 6 while the yellow dot is the position of the sun to the minute. The inner circles indicate the angle of the sun in the sky

Table 9.7 Rock window sun-intersection data

Days from Southern Solstice	Sun angle (elevation°)	Azimuth (compass°)	Time PM (h:mm)	Calendar Date
−49	(39.32) 39.08	(252.12) 252.18	(2:39) 2:40	2-Nov
−48	39.27	251.73	2:39	3-Nov
−47	39.22	251.34	2:39	4-Nov
−46	(39.4) 39.17	(250.9) 250.96	(2:38) 2:39	5-Nov
−45	39.36 (39.12)	250.52 (250.58)	2:38 (2:39)	6-Nov
45	39.43 (39.19)	250.62 (250.68)	3:08 (3:09)	4-Feb
46	39.27	251.06	3:09	5-Feb
47	39.34	251.45	3:09	6-Feb
48	39.41 (39.17)	251.84 (252.89)	3:09 (3:10)	7-Feb
49	39.24 (39.00)	252.29 (252.34)	3:10 (3:11)	8-Feb

Based on the data, the sun most precisely passes through an azimuth of 251° and an inclination of 39.25° on November 3rd and 4th before the southern solstice. Following the solstice, the sun reaches these values again on February 5th, 6th, and 7th as it continues moving toward the equinox.

Table 9.7 helps substantiate the theory of the sun-intersection window, but this is based on the movement of the sun today. Thirteen thousand years ago, the earth's axis was tilted about 0.7 degrees further than it is today (as discussed in Chap. 6). Although the days of the year remained practically the same (365), the sun would have set slightly further south on the day of the solstice. This slight difference may have added a day or two to the days of precision in the table. Currently, day 48 seems most accurate, but it is possible that day 49 was more precise 13,000 years ago. Also, the sun has an angular diameter of 0.5 degrees, and is too bright to look at on cloudless days. Any part of the sun's disk falling inside the gap may have been considered a synchronicity.

Considering all factors, the angle and inclination of the rock gap closely coincides with the position of the sun, supporting the theory that the grid image was used in a way for Paleoindians to foretell the southern solstice and the passage of a new year. Perhaps the vertical lines in the middle of the grid boxes are days when the sun fell behind the side posts of the rock window, casting a shadow. Recall the lack of white paint in the background of the vertical line boxes. Does this represent shadow? It is also quite possible that the rock window itself was more defined than it is today. Perhaps it was a true "window" with a rocky top, allowing just a circle of sunlight to shine through on the 49th day. If so, we should expect that the top had eroded or broken off since then. Regardless, the time frame for precise sun intersection comes down to a single minute on each day, although the sun itself is so bright that direct line of sight without a gnomon or smaller "window" would make direct observation painful.

Today, the "winter" solstice occurs just before the beginning of the rainy period for Monte Alegre. Along with the rains, numerous new streams form, cutting off land routes. Forty-nine days after December 21, much of the land to the west of the hills is transformed from *mato* (tropical savannah) and *cerrado* (woods) into *pantanal* (swamp), and all the ecological resources that transform with them. Assuming that was similarly true 13,000 years ago (which is assumed to be true because these changes are driven by the ITCZ), the "calendar" would have given the Paleoindian foragers plenty of time to move to higher and drier ground from the Amazon River a few hundred meters south and west of *Painel do Pilão*, or to prepare for migratory departure by way of the same river.

Associated Rock Art

Pictographs beneath and to the right of the grid painting add further context that supports a sun theme (panorama available online at http://gigapan.com/gigapans/55312). The representation of circles and dots in relation to horizontal lines possibly reflects or mirrors the sun in relation to landscape perspectives, just as theorized with the sun window. One of the circles below the grid is filled in with an "x" while another is filled in with a "T" design (Fig. 9.12). Both infills resemble symbols filled into the boxes of the grid. The circumscribed "T" is further embedded along a horizontal line that has a rounded cone to its left side. This profile somewhat mirrors the roof of the *Painel* outcrop (visible in Fig. 9.13). Alternatively, the circle drawing perhaps represents the light of the sunrise peering into the crevice beneath the wall. This possibility will be discussed later in the section, *Piercing the*

Fig. 9.12 *Painel do Pilão* linear drawings below grid painting. Raw photo image

Fig. 9.13 Photo of the landscape profile from the "roof" of the *Painel* outcrop to the "sun-capture window" (top). Photo of the sun passing overhead at 10:27 AM on March 21, 2011 (bottom right). Compare the rounded cone shape on the roof of *Painel* to the rounded cone on the roof (bottom left)

Fig. 9.14 *Painel do Pilão* right edge of grid painting—Dstretch_cb_lrd enhanced

Circle with Sun Daggers? Similarly, as the sun progress toward the equinox, it visibly sets further away from the outcrop. This too might be represented to the right of the grid by the dots (with short vertical lines) painted further away from the grid painting (Fig. 9.14).

The wall further to the left of the grid is at a different angle, running ~150–157° southeast and ~330–337° northwest (approximately 153°–333°). Numerous pictographs line this wall and face the east—the direction that the sun rises—roughly 63° to 72° with variations based on local unevenness and inclination. On this wall, the

Fig. 9.15 Painted wall to the left of the grid painting. The "lollipop matrix" image (right edge of photo) faces 66° azimuth—the furthest north that the sun rises on the eastern horizon, which, by definition, occurs on the day of the June 21st Summer solstice. Note the red circular marks on the platform/stage along the bottom edge of the photo)

most peculiar drawing is in the shape of circles connected by vertical and diagonal lines (Fig. 9.15), like the pictographs seen to the right of the grid. These shapes were categorized as lollipops, but they are believed to represent the sun moving through the sky as well. This connected lollipop "matrix" faces 66° east, which is the alignment of sunrise on the day of the northern (summer) solstice. Each day after the northern solstice, the sun rises further to the south, and we can see a progression of pictographs trailing further south. Additionally, the path of the sun as it ascends into the sky curves either slightly to the right in the summer, to the left in the winter, or directly up in the sky on the equinoxes (revisit the sun chart in Fig. 9.11). The lollipop matrix drawing similarly shows the progression of red circles along those lines.

Among the matrix painting are other sky-themed images, including possibly a comet or bolide—the head facing down, toward the sun if the comet were visible before dawn. Also interesting to note is how large it seems (for a comet) and that it is positioned between the equinox matrix and the southern solstice matrix. If these associations really coincided with an unusually large comet visible before dawn (Fig. 9.16), it seems to temporally place that event as occurring about half-way between the September 21 equinox and the December 21 solstice using today's calendar—so late October or early November. As discussed in Chap. 6, the calendar months 13,000 years ago would have been vastly different, but the solstices refer to the position of the sun relative to the earth. This time frame between autumnal equinox and winter solstice coincides again with the Taurid meteor shower.

The ground at the base of the lollipop matrix, on the platform/stage, has some interesting red paint marks, including a partial red footprint (Fig. 9.17, and also visible at the bottom edge of the enhanced image in Fig. 9.15). These marks might serve as important clues to the exact position of a sky observer and/or the placement of a gnomon to cast a sun shadow on the wall. They could indicate how the ancient artists achieved reasonably precise solstice alignments. To the right of my black shoe is a right footprint with a rounded big toe. Near the top of the photo, closer to the

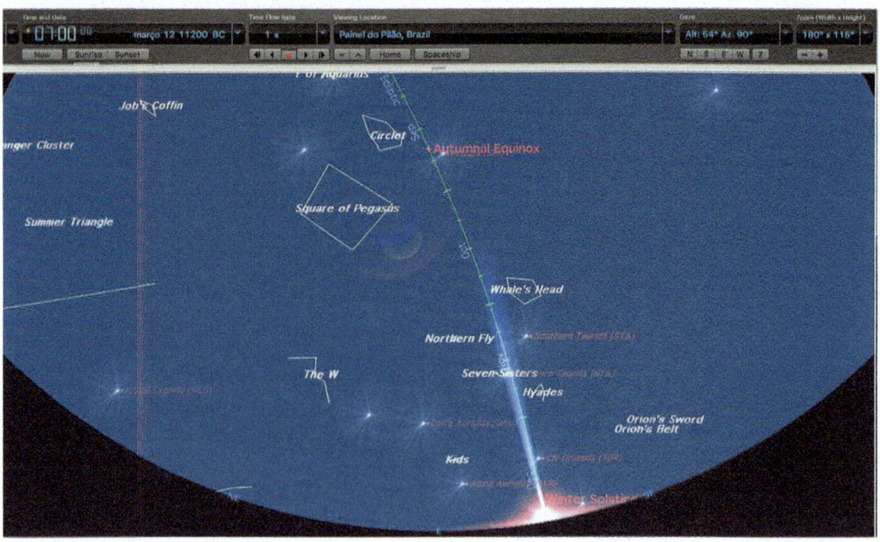

Fig. 9.16 StarryNight recreation of the sky at Painel do Pilão 13,000 years ago, facing east at sunrise, with the lollipop matrix at your back. The path of the sun (ecliptic) is the green line; the autumnal equinox and winter/southern solstice (which occurred around this day, March 12, 11,200 B.C.) are marked in bold red words. Between those two times, along the ecliptic, are the southern and northern Taurids, which occur around November today, but would have occurred around January or February that long ago. Also depicted as a blue streak is the (SOHO P/2008 Y12)—a near-earth asteroid. Alternatively, it could be depicted as the bolide

painted wall with the lollipop matrix, is a red circular mark. If an observer stood on the footprint and held a vertical wooden post at the location of the circular mark, the rising sun would cast a shadow onto the wall, allowing its movement to be tracked.

To the left of the lollipop matrix is a bird-like stick figure with pincer-like hands and feet, and a hook head. Between the lower limbs appears to be a brimmed dot. I do not know the meaning of this figure and I first assumed it was a bird. However, it lacks any sign of wings. It could perhaps be upside down, making the hook head really the tail. It's not possible to decipher every image from so long ago, but the figure is distinct enough to make a note of it. Perhaps it is a therianthrope—a mix of human and animal characteristics. Its significance on this wall, next to the lollipop matrix, could be stellar, so StarryNight astronomy software was consulted for this location (Fig. 9.18). The "sickle" asterism in Leo (representing its head) appears in front of the "hook head" painting before dawn. Instead of a lion, the ancient artist might have envisioned it as this therianthropic figure. I examined the orientation of the sickle asterism across various points in history to understand how its position has changed over thousands of years. In 11,200 B.C.—the earliest period corresponding to radiocarbon dates from the excavation—the asterism is roughly "upright" and visible before dawn slightly south of 90° east. When checking for other radiocarbon dates (4500 B.C., 2200 B.C., and A.D. 1000), the sickle increasingly appears oriented on its back. The 13,000-year-old sky presents a striking match: the sickle is upright and bends in the same direction as the "hook head" painting.

Fig. 9.17 Red right footprint on platform/stage at Painel do Pilão. The footprint is facing the direction of the painted wall. There is also a red circular mark (top of photo) closer to the wall. These marks might have been intentional, demarking the place for an observer to stand and hold a gnomon so that the rising sun could cast a shadow on the wall, recorded with paint

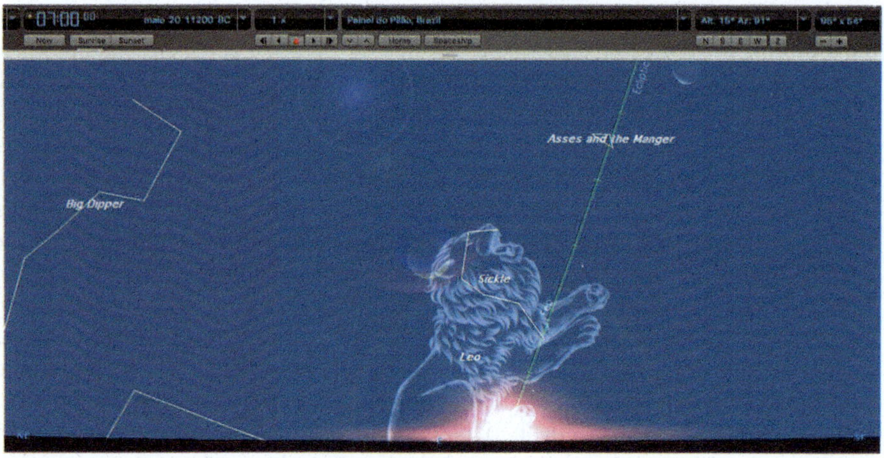

Fig. 9.18 The "Sickle" stellar asterism of the constellation Leo would have appeared in the predawn sky in front of the "hook head" pictograph at Painel do Pilão 13,000 years ago. The sun rising and crossing its midline at the start of each day around the equinox

Adjacent to the pictograph are faint yellow images, challenging to discern without enhancement. Using D-stretch image processing (Fig. 9.15, bottom), I identified two rayed circles, one on either side of the "hook head." These circles may symbolize the sun rising in Leo at dawn (Fig. 9.18), as the constellation crosses the midline at first light. Lower on the wall, yellow pictographs resembling small, child-sized handprints are present, though they appear smudged or blurred, giving them the appearance of yellow dots. The significance of these handprints remains uncertain, but a theory linking them to toddler rites of passage seems plausible. This site may have been chosen for handprint rites in symbolic parallel to the "birth" of the sun each day.

Sliding further south (to the left) of the "hook head" along the *Painel* painted wall brings us to another complex jumble of paintings (Fig. 9.19). I cannot decipher each image, but the series of bent vertical lines and a circle with a plus sign within it might further support the theory that the art generally tracks the rising sun. They are painted on the same wall, and the angle they face is about the same, ~68°. The bent lines may be tallies themselves, or they more specifically might be a record in paint of gnomon shadows as the sun moves from the northern solstice back toward the equinox, and eventually to the southern solstice (moving to the left side of the photo). The vertical lines seem to be erupting from a linear landscape profile. There is a concentric plus sign to the left and embedded on the "horizon" of the landscape line. It may symbolize the sun rising. The sun rises each morning mirroring the sequence of vertical lines, back and forth from northern to southern to northern solstices again and again. Additionally, the boxes positioned lower on the wall are reminiscent of smaller grids. Closer examination of the boxes, as revealed with Dstretch color enhancement (Fig. 9.20), shows several boxes with red points in them, reminiscent of tallies, like the large grid around the edge of the wall.

Fig. 9.19 Painted wall to the left of the "hook head" image (right edge of photo). This cluster of vertical lines and small grid boxes continue further south on the same wall, which only gradually curves, closer toward the southern solstice horizon, which is our winter solstice today, occurring on December 21 at Monte Alegre

Fig. 9.20 Dstretch_lrd red color enhancement of lower portion of wall beneath vertical lines. Note the small grid boxes with multiple small red points in the center of the photo. There are also faint symbols in some of the other grids' boxes

In addition to the ground marks, vertical lines, and boxes, this painted wall has indentations (a horizontal gouge) both above and below the art that look less natural than intentional. Unlike many natural fractures and breaks in the cliff walls of

Monte Alegre, these gouges have very rounded edges. The gouge above the red vertical lines particularly seems to have been made by three or four circular or oval indentations smoothed out until they connected. These gouges carry importance (especially if they were made from circular indentations) because as the sun rises on the days near the northern solstice, direct sunshine will illuminate the gouges for a restricted period of the morning—the low gouge first and then the higher gouge later. Since the sun also changes its angular position as it rises in the sky each day, the use of a gnomon to track daily differences must be made at roughly the same time of day, each day. For a pre-clock (and maybe pre-sundial) culture, the gouges corral gnomon recordings to within one or a few hours. Referring to Fig. 9.11, the sun rises more than 15 degrees high within a single morning hour on the summer solstice (from 6 to 7 for example), and it reaches ~45° altitude by the second hour of daylight. These two gouges, below and above the painted vertical lines, may have been used as observation intervals for reckoning gnomon shadows. Also recall that the incorporation of circular orifices in use with sunshine and shadow has been discussed already at *Serra da Lua*, and the next section discusses their potential use with the upper ledge art here as well (Fig. 9.21).

The wall itself does not gradually curve around, and it is a much shorter distance than the expanse at *Serra da Lua*. All the paintings in this panel are oriented to within a few degrees (~68° to ~60° east), making them all generally face the northern solstice sunrise. The red-painted vertical lines could represent the change of sunrise positions at a pace of one or a few days, but still the use of a gnomon would have been necessary to distinguish slight angular changes. Interestingly, there is paint on the platform, which possibly indicated locations to place a gnomon so that sun shadows could be recorded. Any gnomon (which is usually just a wooden stick) has long since decayed, but the red points visible on the stage (revisit bottom of Fig. 9.15) may be proof of such a practice. Notice that the red ground points are just in front of the lollipop matrix, the vertical lines, and the large grid box. They were not dripped haphazardly, nor do they line the entire floor right beneath the wall, where most paint would land if unintentional. There is also no sign of yellow paint drips and drops. If these were haphazard, why would red paint be slopier than yellow paint? If a gnomon were this close to the wall, casting a shadow as the sun rose each day, the vertical shadow cast by the gnomon would progress from left to right in the photo—as the sun rises slightly to the south each day from the northern solstice to the southern solstice. But where is the rock art record of sunrise at the southern (winter) solstice? For that we turn to the upper ledge art.

Upper Ledge Art

The upper ledge rock art at *Painel do Pilão* begins on the east side with an opening large enough that it can be considered a small shelter (panorama available online at http://gigapan.com/gigapans/191127). Although the shelter is not taller than 1 to 1.5 meters high, it is wide enough for several people to sit or lay down comfortably. The

Fig. 9.21 Closeup of the higher gouge (top photo) above the red-painted vertical lines and the lower gouge (bottom photo) beneath the pictographs. Both gouges have uneven linear edges that are nonetheless rounded, not angular as expected from natural breaks in the rock. The left corner of the bottom photo is also where the nearly filled-in subterranean cave entrance is located. Long ago, this cave entrance would have tunneled through to the excavated rock shelter on the other side

shelter is currently perched over 3 or 4 meters above the natural ground. The natural ground underneath the perch is a mostly filled-in entrance to a small cave underneath the outcrop, so the natural ground was much lower in the past, making the perch even higher to reach.

Entering the perch from the eastern direction would not have been possible 13,000 years ago (and still is not) without climbing a vine or tree root (a vine or root visibly hangs below the opening in Fig. 9.7) or using a human-made rope or scaffolding. A difficult rock climb was perhaps possible from the platform/stage, scaling up the lower ledge rock art, and then scampering across smooth rock while moving southward; however, there are no signs of repeated climbing marks or erosion. The gouges, and several other orifices on the painted wall, could have been perhaps used as foot or hand holds—making the rock climb a bit easier—but scaling sheer rock would have been generally easier going from south to north, moving from the perch not toward it.

Painted inside the perch is a ~10 cm diameter red dot that faces outward to the eastern sky (Fig. 9.22). This red dot is angled a bit higher in the sky though, not angled at the horizon. Peculiarly, three circular orifices lay above the dot, the middle orifice being positioned right behind it. A direct line piercing the center of the red dot from the central orifice aligns to ~114°—the path of the rising sun on the day of the southern solstice. All three orifices appear to be angled in the same direction (~114°), but their angular separation to each other from the center of the red dot is about 12°, or ~24° total (drawing a line from the left orifice through the center of the red dot would point to ~102°, and the right would be ~126°). For clarity, the orifices don't align to those directions; they align roughly parallel to each other. But perhaps the orifices were used to rectify the red dot. This red dot begins the sequence of upper ledge art, which also appears to record the passage of the sun, and the number of shallow circular orifices alongside the art is numerous.

Fig. 9.22 Small shelter "perch" at the southeastern edge of the outcrop, located high, at difficult accessibility. The shelter pans wide from roughly 80°east to roughly 260°west, but the red disc at the right edge faces ~114°, the position that the sun rises on the day of the southern (winter) solstice

The northern corner of this perch also has red paint resembling rays on the outside rim, and a stick figure on its side also appears to be leaving the opening (Fig. 9.23). The stick figure resembles either a humanoid or perhaps a frog, given its curvy limbs. A similar stick figure is placed a few centimeters further along the rock, followed by a third one a bit further away, creating the impression of motion, as if it's the same person or frog climbing or swimming across the rock. These "rock climber" pictographs were painted on different sections of the rock, each facing a different cardinal direction. However, taking direct compass measurements of the orientation of most of the upper ledge art was not feasible, as there was insufficient ledge space to stand beneath them or to precisely determine their angle relative to the sky.

Fig. 9.23 Red rays emanating from north corner outside rim of the "perch" along with the first rock climber stick figure (top photo)—note the very shallow circular depression inside the right edge of the perch. All three rock climber pictographs in sequence moving from south to north across the upper ledge (bottom photo)—similarly note the numerous small circular orifices and oval gouges

Most of the art on the upper ledge faces higher angles in the sky, between 30° and 60° inclinations, making cardinal directions somewhat variable. For example, the third rock climber pictograph had an indirect compass measurement of ~120°. Consulting the SunEarth chart in Fig. 9.24 shows that the sun reaches a relative position of ~120° southeast around 8:36 AM on December 21st today, when it is about 42° high in the sky. However, the sun also reaches 120° southeast at approximately 10:50 AM on February 21st, when it is nearly 75° high in the sky. While the sun passes through this cardinal direction every day in between, it does so slightly later in the morning and at a progressively higher position in the sky. Without a reliable anchor to determine the exact inclination alongside the cardinal orientation, it is not possible to pinpoint a specific day.

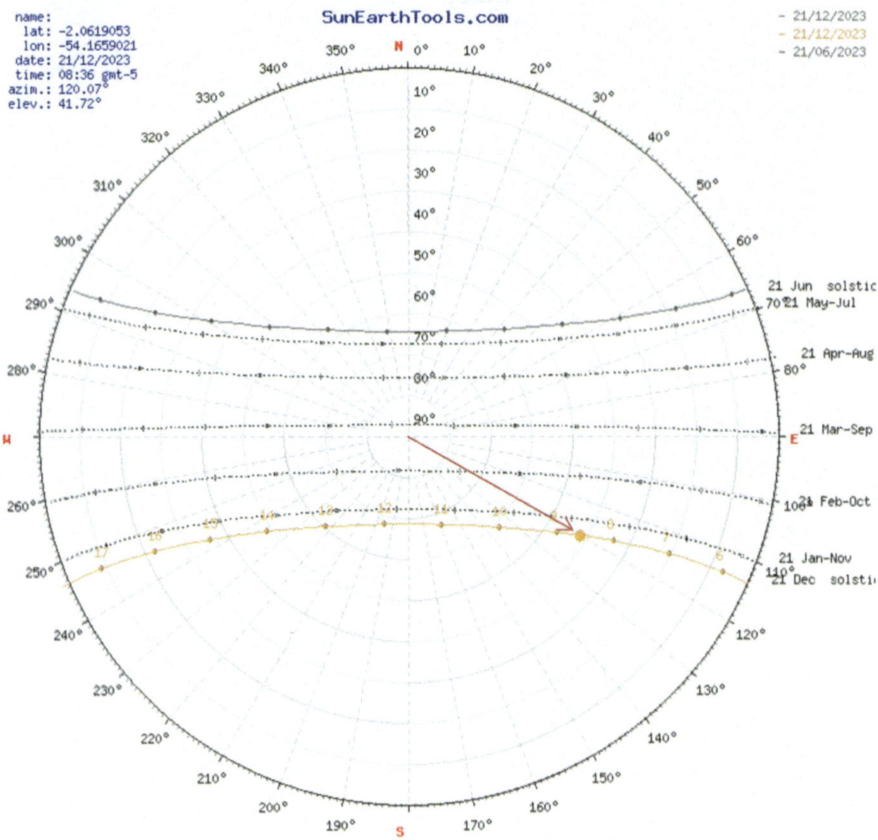

Fig. 9.24 SunEarthTools.com diagram of the sun at 120° on December 21st at *Painel do Pilão* as seen relatively from a bird's-eye perspective looking down. With the observer at the center of the diagram, the sun would be seen at an elevation (inclination) of 41.72°. However, the sun would be seen at 120° everyday along that red line, from December 21 to nearly March 21, but at a higher elevation (inclination) each day

The orifices located near the art may have served as such inclination anchors. While many of the orifices might be naturally formed, some elongated oval gouges and deeper orifices suggest possible human modification. Additionally, several orifices are arranged parallel to one another. If all these features were solely the result of erosion, they would likely exhibit more uniform shapes and depths. Unfortunately, during the years I conducted research and measurements at the site, I did not recognize the potential of these orifices being intentionally altered to capture sunlight at specific inclinations. As a result, I do not have a detailed record of their inclination and orientation. Much of this section, therefore, is based on qualitative evidence and descriptive observations.

Following the rock climbers are two more smaller red dots, one with a small line pointing to two lollipop motifs angled to ~53° northeast, estimated from compass measurements taken far below (Fig. 9.25). The sun reaches this direction but not anytime during the period from winter solstice to equinox. Instead, the sun appears to be in this direction at ~10 AM on the day of the summer solstice—but also later and later, and higher and higher in the sky, each morning until the days near the equinox. As mentioned previously, a specific inclination would need to be known to pinpoint the exact day; however, this association of the lollipop motif does strengthen the idea that this motif is more of a reference to the summer sun's motion throughout the day. The pictograph above the lollipops is abstract, but there is a handprint to the right of the lower lollipop. The handprint appears to be that of a left hand overlaid by a red dot. Another red dot is part of a larger composition above the handprint, and color enhancement (Fig. 9.26) reveals that yellow pigment composed part

Fig. 9.25 Upper ledge lollipops, estimated to face ~53°. The "stick" indicates a "rise" higher in the sky, if symbolic of the sun, continuing the implied "motion" of the rock climbers

Fig. 9.26 The Abstract composition with scale and Dstretch_cb_lre color-balanced red enhancement reveals some yellow pigment underneath a red dot. The bent line on the right side and two lines beneath appear almost avian. But other than the red dot, this image is largely undecipherable. The left handprint and dot are more visible left of the scale

of what seemed to be a rayed dot. The composition is currently undecipherable, but an arc and the lines projecting to the right appear to continue the "motion" across the rock.

Some curvilinear lines and two bizarrely drawn dots with elongated, finger-like rays continue the "motion" of the art past another lollipop drawn at a diagonal and facing higher in the sky, perhaps to indicate a late morning position of the sun (Fig. 9.27). To the right of this is what appears to be the profile of a horned (bovid?) animal, although the "head" is separated from the profile of the torso, and the head appears to be composed of a large lollipop motif with horns. This part of the wall does curve gradually, running from about 60° to ~18° (the diagonal lollipop) to just about true north (the bovid). Beneath the bovid is another stick figure with pincher-like limbs held out to the side. The torso is relatively short compared to the limbs, so at first it seemed to represent a crab. Using Dstretch photo enhancement allows closer inspection, though, revealing that it is the "hook head" figure on its side, with the head still curving to the body's right side. This is exactly what happens when it reaches the zenith of the night sky. Naked-eye sky watchers knew which constellation reached zenith at midnight because it was the constellation that became visible on the eastern horizon just after sunset in the west. According to StarryNight software (Fig. 9.28), if the hook head pictograph represents the sickle asterism of the constellation Leo, as introduced previously with Fig. 9.18, the asterism reached the zenith of the sky at midnight around November 16, 11,200 B.C., directly overhead, which was a month before the vernal (Spring) equinox. Perhaps the painted bovine

Fig. 9.27 Bovine and sideways "hook head" (top photo). There is also an upright, "vertical head" pictograph on the right edge. Note the faded white background paint. Dstretch_rgb0 red color enhancement (bottom photo) makes the hook head more visible

reflects another constellation the ancient artist(s) envisioned from the combination of the Diamond of Virgo and Kite asterisms. Likewise, the upright stick figure to the right may have been envisioned from a composite of the big and little dippers.

We can't know with certainty how people so long ago envisioned the stars at night, but the hook head, and its changed position, offers a compelling consideration. Evidence for sun tracking is more robust though, and it continues with the upper ledge art. The next pictographs along the ledge are found on its floor, facing up toward the zenith of the sky.

There are two red-painted dots in progression from east (right corner) to west (center) tracking the movement of the sun from late morning to noon. The red dot in the right corner has faint remnants of what may have been a spoked concentric circle, when viewed with Dstretch_cb_yub color-balanced general enhancement (Fig. 9.29). The larger red dot in the center of the photo has still visible rays emanating from it. The inclination and cardinal orientation are not precise enough to give

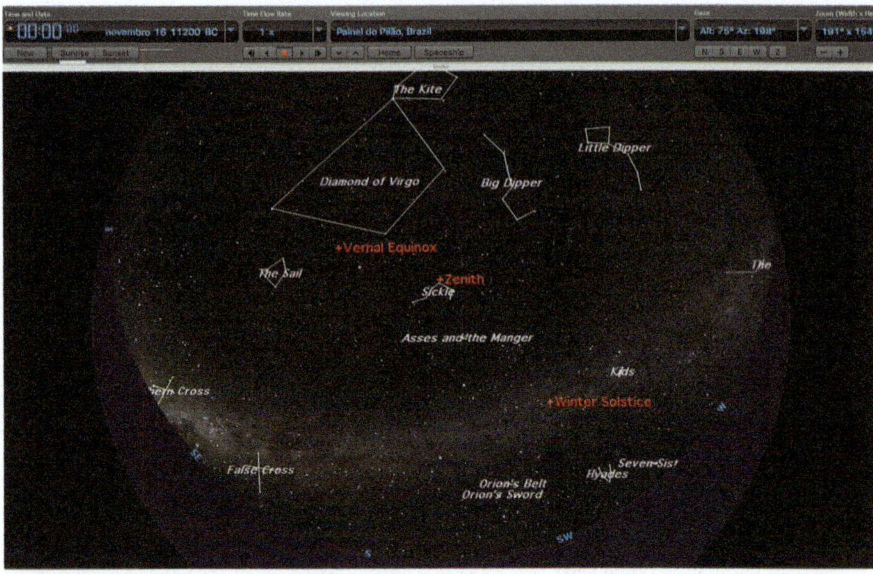

Fig. 9.28 StarryNight software recreation of the night sky at *Painel do Pilão* on November 16, 11,200 B.C. (13,000 years ago) at midnight and facing 198° south. This was about a month before the vernal equinox, which is equivalent to occurring today near February 21, the last month of winter. The sickle asterism is on its side as it passes the zenith

Fig. 9.29 Two red-painted dots in progression from east (right corner) to west (center), as indicated by the big white arrow, track the movement of the sun from late morning to noon. The faint red, yellow, and white paint remnants are made more visible with Dstretch_cb_yub color-balanced general enhancement. The faint purple meandering scrawls are undecipherable

an exact date, but the time frame for the sun to pass over these paintings would have been near the equinox. There is also a third red, paint- chipped dot to the left (south) of the other two. The sun would have passed directly over this third dot days to weeks earlier (or later), as the year progressed from or toward the winter (southern) solstice.

Further back against the ledge, much of the floor is still covered in a white background paint, but there is a rounded knob on the floor that has the white background mostly worn away. Painted along the curvature of the knob is a red-painted, elongated oval divided into grid-like sections (Fig. 9.30). It is hard to determine the number of sections because there are vertical, horizontal, and oval dividing lines, but there are at least 16 sections generally arranged linearly. The elongated oval runs nearly true east–west overall, although it's painted around a curve. This image

Fig. 9.30 Elongated grid painted on curved knob on the floor of the upper ledge runs roughly east–west, facing out at nearly 0° north. In the normal-color photo (top), note the rounded knob is the only place where the white background wore off. In the Dstretch_cb_lrd red color-balanced enhancement (bottom) a scale is visible, as is the possible "gnomon" point

possibly provides more evidence that grids were used in conjunctions with gnomons. About 7 cm above the grid is a red point, perhaps indicating where to place a gnomon. The significance of doing so may have been so that as the sun crosses the zenith each day around noon during winter, it will cast a shadow that intersects this elongated oval near midday. However, as the days near the equinox, the shadow of the gnomon will get shorter each day, until it no longer reaches the pictograph—the day when a gnomon does not cast a shadow as the sun passes overhead is roughly the day of the equinox (actually a few days away, since Monte Alegre is 2 degrees south of the equator). The only real problem to this theory is that the roof of the outcrop projects over the upper ledge.

Sunlight does not directly shine on most of the upper ledge until "summer" time.

Ethnographic accounts of Amazonian Indians suggest that more importance was given to the day of the year when the sun cast no shadow from a gnomon at midday (Aveni and Hartung 1981:52; Krupp 2012:316–317). Such an event only occurs for just one day of the year near midday when the sun passes through the exact zenith of the sky for a given location that must lie near the equator at latitudes between the tropics. Monte Alegre, at 2°3' south, meets this criterion. However, the day this occurs is not on the day of the equinox.

Today, this day of virtually "no shadow" occurs on March 15 and September 28, about seven days away from the equinox. On these days, at about 12:28 PM, the sun reaches the exact zenith at 89.9° altitude (determined with StarryNight software for the year AD 2010). On those days, however, the sun rises at the horizon at ~92°, not 90°. Even 13,000 years ago, according to StarryNight's reconstructed ancient sky, the sun would have rose in the morning at 91.48° local azimuth (accounting for obliquity and precession) on the day it passed exactly through the zenith, thereby casting "no shadows" at the zenith. Unfortunately, much of the outcrop remained shaded (from the overhanging "roof"), even on March 21, 2011, when I tracked the progression of the sun through the day (Fig. 9.31). Recorded video still images show sunrays only shining on the back of the east wall and a single ray of sun hitting the west corner of the main grid wall at the spot not covered in green lichen at ~10:30 AM. At 4:30 PM, the sun has passed to the west, behind the main grid wall, where another rocky outcrop further blocks the area from direct sunshine. The sun still must move further north (summer), past the "lip" of the roof to illuminate the upper and lower painted ledges.

The next pictograph along the upper ledge provides the strongest evidence that a gnomon was used at *Painel do Pilão* to track the sun. Another red point lies just (northeast) beneath a red spiral motif, and a red rectangular bar lies just above (southwest) it (Fig. 9.32). An imaginary line connecting the point, and the bar would intersect the spiral. Such a line would become visible if a gnomon were held vertical at the spot of the point so that the sun passing by in the midday sky casts a gnomon shadow. The direction of the imaginary line runs roughly northeast–southwest ~50° and ~230°. However, like many of the other upper ledge alignments, the inclination of the sun would need to be known to pinpoint the intended day and time that is referenced. Currently, it can only be narrowed down to summer mornings. If the gnomon were a specific length, perhaps the shadow that was cast terminates at the

Fig. 9.31 *Painel do Pilão* still images from videos recorded on March 21, 2011, at ~10:30 AM with sun illuminating the east wall (top left), and just the corner tip of the main grid wall at the spot at the center and not covered by the green lichen (top right). At ~4:30 PM looking west across the main grid wall (bottom left) and looking up at the "roof" from the main grid painting (bottom right)

Fig. 9.32 Spiral and shadow marker. The red point beneath and red rectangle above the spiral, visible in the Dstretch_cb_lre red color-balanced enhancement, further support that a gnomon may have been placed on the point so that the sun cast its shadow through the spiral and rectangle. The angle runs between ~50°and ~230°—with the late morning sun casting gnomon shadows that intersect the spiral on days near the summer solstice

Fig. 9.33 Small, red-painted rectangular grid boxes along the rim of gouged orifices on upper ledge. This set of grid boxes are more rectangular and there is no gnomon point nearby. However, the grid borders the rim above some elongated gouges (left photo), and the grid is more visible above the right edge (top right) with Dstretch_cb_lre red color balance

edge of the rectangular bar, making the sun's incline very high—possibly on the day of the summer solstice. However, if the bar is simply a connecting point, a longer shadow would pierce the spiral and bar each morning at ever-earlier times of the day until the sun approaches the equinox again. It is interesting to note that following the imaginary intersection line further north behind the gnomon point results in that imaginary line also intersecting one of the lollipop motifs on the lower-level paintings to the right side of the main grid (revisit Fig. 9.7).

Deeper back inside the upper ledge is a linear series of elongated gouges that create a thin rim that was painted with narrow rectangular red grids (Fig. 9.33). There is no "gnomon point" near these grid boxes. However, the grid boxes have red points in them, and the gouges show signs of having been altered. As was stated previously, the angle of inclination for the sun would need to be indicated in addition to the cardinal direction the orifices face. The orifices generally face north, but the inclinations of the gouges are unknown. They would be, however, filled by the rays of the sun briefly on summer days around midday. The red points in the boxes seem to represent a tally of days once again. It is uncertain whether the white line intersecting a grid box was painted long ago to trace a shadow.

Following these grid boxes, the art on the floor changes again to some curvilinear abstract designs (Fig. 9.34). Part of the design resembles the upper torso of a humanoid (perhaps), and it contains a blend of red and yellow pigment. There is also a three-pronged symbol that bears some resemblance to written script one might find in East Asian language. I have not found any significant connection, however, other than coincidental similarity of style. I do not have any theory for what any portion of this upper ledge rock art symbolizes, other than perhaps more constellations. They do face the northern sky, close to true north, but like the "hook head" motif, the design might be a reference to stellar asterisms along the zodiac that pass through the zenith of the sky, or they might reference other constellations that are more formally seen high in the northern sky.

Fig. 9.34 Abstract curvilinear scrawl with humanoid torso directly facing the zenith of the sky (above) as well as the north sky (top photo). Dstretch_cb_lre red color-balanced photo enhancement details the use of separate red and yellow paint, not orange paint (bottom left) and a three-pronged symbol (bottom right) reminiscent of a logographic script character. Art on the upper ledge continues from the ground to the wall and ceiling from here

There are still consistent repetitions of orifices, constellations, and sun shadows present. For example, three parallel orifices are positioned underneath the abstract curvilinear scrawl, alongside a large horizontal gouge in the wall beneath the wall/ceiling art. Below the gouge, a small circular orifice bordered in red paint can be found, with another painted orifice located to the right of the gouge. The orifice beneath the gouge is angled toward a high position in the sky, while the other painted orifice aligns with the northeastern path of the sun.

Art on the ceiling includes a "kite" motif with handprints, an avian-like pictograph, and another multi-pronged logograph (Fig. 9.35 top). Additionally, a long, snake-like image is present. These images could potentially represent stellar asterisms along or near the zodiac between the equinox and summer skies, though this is speculative. The kite pictograph might represent the "kite" asterism in Boötes,

Fig. 9.35 Wall and ceiling rock art on the upper-level ledge. Note the three orifices underneath the abstract curvilinear scrawl, and the large gouge underneath the wall/ceiling art. Underneath the gouge is a small circular orifice bordered in red paint. The ceiling also has a "kite" with handprints, an avian-looking pictograph, and another multi-pronged logograph

mentioned previously, or possibly the "great diamond" asterism in Virgo—though it is uncertain what object or concept people 13,000 years ago might have associated with the diamond shape. The avian-like pictograph could perhaps correspond to the "teapot" asterism in Sagittarius, while the snake-like pictograph bears some resemblance to the "W" shape of Cassiopeia.

Finally, it was mentioned at the beginning of this section that the upper ledge is easier to reach from the north than to scale the "perch" from the south. In fact, the hill behind the outcrop does rise, and the upper ledge can be easily reached (after

Fig. 9.36 Orifices on the southeast side of the *Painel do Pilão* outcrop (left photo)—the red lines face ~132° southeast, but the lowest orifice faces ~114°, the position of the southern (winter) solstice, and the other orifices face the risen sun in the eastern sky. The author reaching the upper ledge from a western circuitous route (right photo)—the numerous orifices here were not measured but generally face the direction of the setting sun around the summer (northern) solstice

walking circuitously around) from the west (Fig. 9.36 right). There are numerous circular orifices facing this direction as well, which is the direction of the setting sun in late afternoons near the northern (summer) solstice. The far east side of the outcrop was a previous source of error in measuring angles. The cardinal direction of the red-painted lines was not calibrated to true north. Returning to verify these measurements in 2016, and again in 2023, the lines truly face 132° southeast calibrating to true north, but the lowest orifice above the center of the red lines indeed calibrates to 114°, the position of the rising sun on the day of the southern (winter) solstice. Note that the other orifices face roughly parallel but are angled at a higher inclination, possibly to capture the sun progress through the morning. Note where these orifices appear and where they do not.

Looking at the entire outcrop, orifices are found from east to near zenith to west—all along the ecliptic of the sky, possibly spanning through parts of both summer and winter, and from morning to evening. If every orifice formed naturally, why would they be clustered on some walls and almost non-existent on other walls? The main grid wall barely has any circular orifices like those on the upper ledge. The orifices appear to complement the circular pictographs at most places on the outcrop that receive direct sun rays. Perhaps some are natural, but there are also a few orifices—and other circular indentations—with varying depths. It seems logical that if some (or most) orifices were hollowed out intentionally by ancient sky watchers to capture sunlight at specific intervals of the day and year, a few were made or shaped in error. We might be looking at multiple, honeycomb-like orifices as the result of trial and error. Just as I've made sun measurement mistakes in the past, perhaps so did the ancient sun-tracking artist(s).

Overall, by examining the upper ledge of paintings, the theory that the ancient inhabitants kept track of the sun not just through the year—but also partly through the day— is further strengthened. Circumstantial evidence suggests they used gnomon at specific positions, indicated by red points painted on the ground, so that sunshine would produce a gnomon shadow across painted grids and at least one spiral. They also appear to have created or reshaped orifices into gouges and circles that became illuminated only when the sun reached a certain inclination in the sky (thereby tracking the progression of daylight). There are ethnographic references to traditional Amazonian cultures keeping track of different stages of daylight.

In the 1935 book *O Selvagem*, Amazonian scholar José Vieira Couto de Magalhães detailed how the Tupí Amazonian natives divided their daily activities into two-to-three-hour intervals based on the position and color of the sun in the sky during the day. These intervals are specifically named and correspond to positions in the sky that we define as *coema* = dawn, *coarací iauate* = meridian, *caie ou iandára* = noon and is a literal reference to reaching the tip or peak of the sky, *a'ra* = afternoon and relates to birds that take flight, *carúca* = evening until dusk, *pitúna* = night, *piçaié* = midnight, *pitúna pucú* = graveyard shift, *coema pirãnga* = dawn and literally translates as "red of the morning" (Magalhães 1935:77–78). Other scholars noted that the Tupí similarly kept track of the position of the moon, Venus (called *iacitatá-uaçu*), the Pleiades (called *ceiuci*), and other stars at night (Lima and Figueirôa 2010:307,308). Magalhães further explained that some intervals, especially of the night, were named by the dominant animal noises heard during that time frame (Lima and Figueirôa 2010), much like naming dawn after the crowing rooster, or naming midnight after the nightingale, screech owl, or whippoorwill. Stephen Hugh Jones (1982) learned that the Barasana relate the setting of the Pleiades to the return of the bobolinks bird (*Dolichonyx oryzivorus*), which flock to South America in the fall (López-Lanús et al. 2007), and they relate the setting of Orion's belt and sword to the return of a flock of small black birds (Kelley and Milone 2011:469). In this way they associate the seasonal motion of stars and constellations with important ecological patterns in nature (Kelley and Milone 2011:469–470).

We can therefore imagine that "prominent" nighttime animals may have been assigned the zenith constellation for the seasonal period that they make noises. It wasn't that the stars clearly *looked* like that animal, but instead was associated in whatever way possible to serve as a mnemonic record of the animal—the entire zodiac signifying ecological cycles throughout the year. Such specific details for periods of the day and night, based on the position and color of the sun, and other celestial sightings, indicate a possibly inherited tradition of observing the subtleties of the sky because their motions would have required meticulous records extending deep back in time.

However, the evidence for sun tracking at *Painel do Pilão* is not yet finished. There is one more area of the outcrop that may have been altered for sun-tracking purposes. So far, the chapter has focused on rock art and alterations above the platform/stage, open to the expansive sky. Next, we return to the semi-subterranean shelter underneath the *Painel* outcrop—to the excavation pit and those diagonal gouges in the wall.

Piercing the Circle with Sun Daggers?

Another sun alignment perhaps occurs with a single red pictograph discovered above the excavation unit on a narrow under hang from the ceiling (Fig. 9.37). The under hang shows signs of having been altered intentionally, and its exact orientation has similarly given me great difficulty to pinpoint. The arrangement of the painted red circle will be described in detail because it was the greatest source of error in a previous retracted journal article. But, before revisiting the data and identifying the errors, let's restate from the introductory chapter that the significance of the rock art at Monte Alegre is not *the degree* to which it aligns to the sun and stars. It's that it appears to be a record *attempting* to track the sun and stars.

Problems with previously published alignments of this underside circle arise from several compass measurements that yielded inconsistent readings, ranging from 90° to 109°, 113°, and even 120°. It was later suspected that compass measurements taken closer to the rock may have been influenced by higher-than-normal elemental iron concentrations. The durable platform shelter discussed in the excavation chapter is the likeliest source of this potential iron interference.

Another issue relates to the angles of inclination associated with this painted circle. Because it faces west—within the part of the shelter fully enclosed from the sky—its alignment with the sun is indirect. If the circle symbolizes the sun, it is not oriented toward it. Directly beneath the circle is a triangular opening that permits views to the east. The triangle widens toward the left (north), allowing some visibility of the sky. However, the area directly below the red circle offers a limited view, from the horizon to only about 25° above it.

Fig. 9.37 Photo of the under-hang painted red circle above the excavated shelter at *Painel do Pilão*. The red circle is right of center, near the top edge. The platform/stage is the wall supporting the toolbox, creating a triangular window with the overhang. Through that triangle, the northeast horizon, with the distant Amazon River, can be seen. The author is at the left edge, holding a camera down at the excavation floor before excavation began, on April 27, 2011, at 10:09 AM. Also note the faint white background paint on the under hang and wall

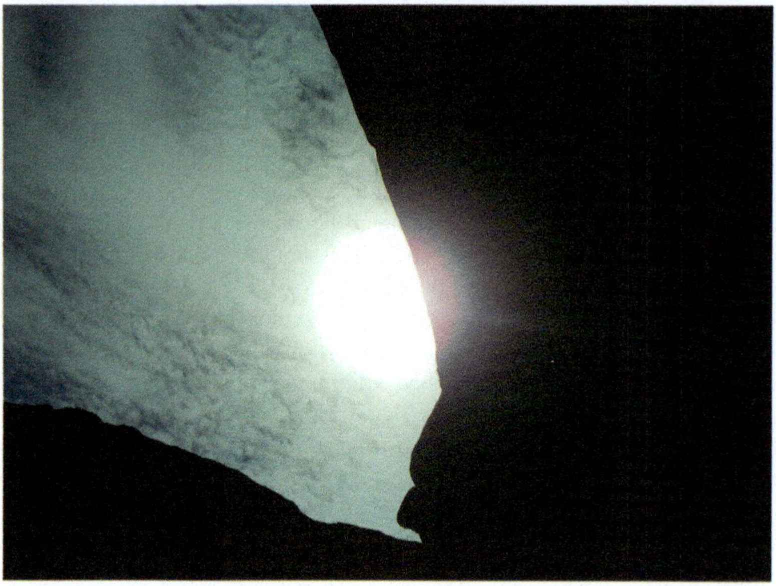

Fig. 9.38 Sun shining to the left of the under-hang red-painted circle above the excavated shelter on May 9th at ~9:38 AM before level 7 was excavated. The cardinal direction measured with a compass was 75°; correcting to true north makes it 57°

A third issue, previously overlooked, concerns the point of observation—specifically, where an observer would have stood to take measurements or view alignments. Approximately 13,000 years ago, the ground level within the shelter was nearly 2 meters lower than it is today. From that lower position, an observer would not have been able to see the horizon at all.

The first time I took special note to directly measure the red circle was during my excavation of the shelter, after digging down to level 6. Even at that depth, I was no longer high enough to look out of the triangular "window" from the bottom of level 6 (Fig. 9.38). Later compass measurements were taken after I backfilled the excavation (current ground level). And more effort was not spent considering where an observer would have stood either. Some compass measurements were taken while standing closer to the platform/stage, while others were taken from the center of the shelter. Perhaps the ancient artist(s) who painted the circle knew where to stand without the need to mark it, but the depth of the shelter would have been a problem regardless. Later measurements in 2016 and 2023 were taken by crouching down atop the platform. The circle (Fig. 9.39) could have been painted long ago by an artist crouching or laying down on the platform, but to make repeated visual sightings from within the shelter, either some sort of ladder was used, or an indirect observation was made. One that did not require directly looking through the triangular window.

After checking and double-checking the circle in 2016 and 2023, it seems that the under hang runs southeast–northwest (154°–334°). This cardinal direction was taken by placing sticks parallel to the under hang so that they extend on top of the

Fig. 9.39 Underside red circle photo taken on December 20, 2016, at 5:09 PM (left) from atop the platform stage (sun was setting in the west behind the camera). Photo taken on August 19, 2023, at 10:54 AM (right) from atop the platform stage (sun in front but above the under hang). Note that the horizon is washed out due to contrast in the left photo, but visible in the right photo. Also note the vegetative growth

platform stage, outside of the shelter. Then, standing above the shelter and positioning the compass in line with the sticks kept the compass far enough away from interference with the rock. Assuming the painted circle runs perpendicular to the section it is painted on, the circle faces 244°, and ~64° behind it. So, the triangular window beneath the circle should perhaps align with ~64°. Additionally, the inclination angle for the horizon is only 0° to ~25° high for the section underneath the circle. From this new determination, it seemed at first that the angle was too far north, since the northern solstice is 66.5° today (and about 65.8° 13,000 years ago).

However, SunEarthTools.com (2024) helped verify that as the sun rises in the summer, it moves further north, crossing the cardinal direction of 64°, near the inclination of ~20° at ~ 7:12 AM on June 21st and its surrounding days today (Fig. 9.40). Thirteen-thousand years ago, this also would have happened on the days around the northern solstice—just a bit earlier in the morning. The diagram is not drawn to scale but is meant to convey the movement of the sun at *Painel do Pilão* in relation to the contour of the main outcrop with the platform and the triangle window containing the painted red circle. Note that afternoons when the sun is north of the equinox allow the sun to shine a "dagger" into the shelter. Days closer to the equinox should produce a narrower dagger. Days near the summer solstice should produce wide sun daggers.

While this suggests just one precise angle, the problem remains how this would have been observed. Not only would the triangular window have been over 3 meters above the ancient shelter surface, but the sun's blinding light makes any direct sighting near impossible. If the intersection happened closer to dawn, as suggested around 6 to 6:30 AM, the sun would have been much less intense, but a cloudless dawn would still have likely filled the entire window with bright light (Fig. 9.41).

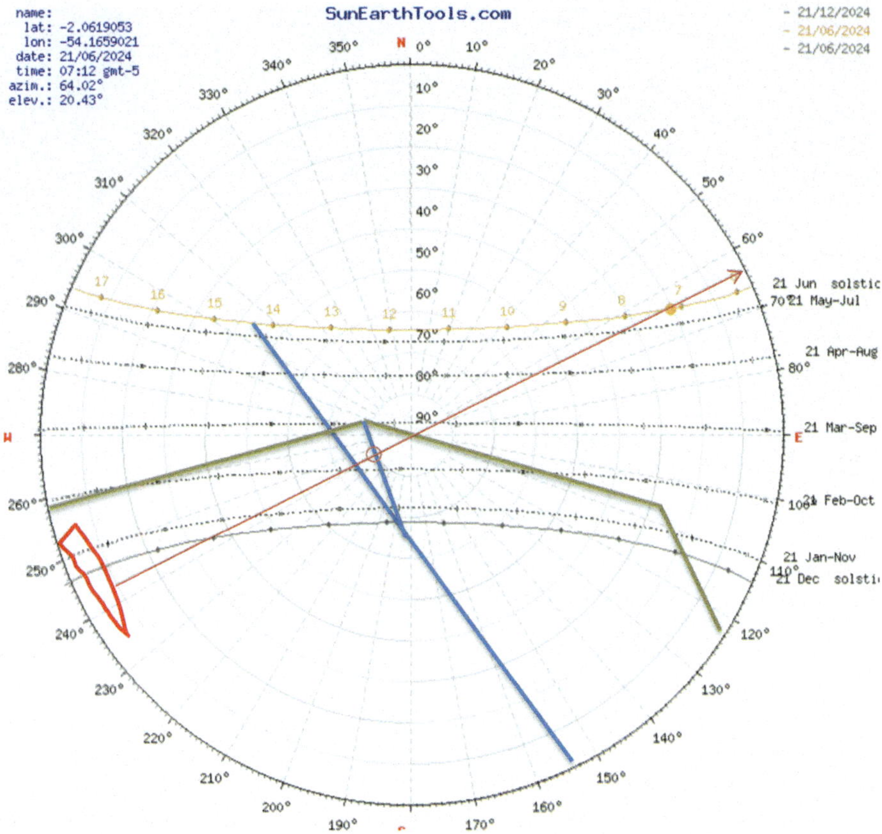

Fig. 9.40 SunEarthTool displaying the position of the sun on the morning of June 21, 2024, around 7:12 AM. Though not to be interpreted as this precise, the diagram exhibits the movement of the sun at *Painel do Pilão* in relation to the contour of the main outcrop and lower-level paintings (green line), the platform stage (long blue line), and the triangle window (short blue line) with the red under-hang circle. The red triangle represents a sun dagger effect

Fig. 9.41 Triangular window indirectly photographed on April 26, 2011, at 9:52 AM while the author was measuring depth in the excavation pit with a local assistant (cropped out for anonymity). Note the radiant sunlight shining through the triangle. The red circle is just off camera but would have been positioned above the visible roof. Compare with Fig. 9.36

And unlike the red points along the outcrop surface, there are no indications where a gnomon might have been placed. Perhaps the illumination of the sun shining against the back (west) wall of the shelter space was enough indication for the ancient inhabitants. But considering the detailed effort that much of the rest of *Painel do Pilão* seems to exhibit, some sort of artistic indication should have been found in that part of the shelter, and yet no trace was discovered. There is one thing in the shelter, though, that I had not previously considered—until sorting through photos in preparation for this book.

During the days of excavation, from April 27th to May 15th, numerous photos were taken inside the shelter. On cloudless days the sun would shine intensely into the shelter, which was a problem for taking level and artifact photos. When possible, direct sunshine was blocked to take better photos. However, some photos and video recorded the sun in the shelter, and it only became clear while reviewing photos years later that the triangular window produces a sun dagger across the shelter (Fig. 9.42). This occurred in afternoons as the days progressed from April (Fig. 9.43) to mid-May (Fig. 9.44) during the approximately two weeks of excavation. As the afternoons progressed, the sun daggers also appear to inch toward (Fig. 9.45) the fully roofed "cave" east inside the shelter (Fig. 9.46). Some of the sun angles parallel the gouges that remain in the platform wall (revisit platform/wall gouges in Chap. 8, panorama also available online at http://gigapan.com/gigapans/81492). Sometime after the vernal equinox, the sun is far enough north that it continues shining light narrowly through the triangular window. The light that appears high in the west corner of the shelter, underneath the red circle, slides down and across the shelter as the afternoon progresses. As the days continue, the sun continues sliding

Fig. 9.42 Sun dagger near center of shelter ground. Photo taken May 14 at 12:05 PM

Fig. 9.43 Sun across platform wall. Still images of video recorded on April 27 at 2:59 PM seconds apart. Author pictured in photo

Fig. 9.44 Sun across platform wall and unit. Photo taken on May 10 at 2:34 PM. Author pictured in photo

Fig. 9.45 Sun dagger across platform wall behind the author's ear in the top right corner. Some of the wall gouges are faintly visible to the left of the author's elbow. Still image of video recorded on April 27 about 3:05 PM

Fig. 9.46 Shelter cave (filled in) looking clear through to the other side. Still image of video recorded on May 14 about 11:51 AM

further north and the sun dagger gets wider each day, as more sunlight shines through the triangle window.

Unfortunately, because I never arrived at the site in the earliest few hours of daylight, and because the excavation was just two weeks, I do not have a long enough record to further substantiate the sun dagger theory. Judging by complex angles of the rock and the dynamic movements of the sun, the edge of the excavation ceiling (which is the bottom of the main grid wall) may have been "contoured" by the ancient artist(s) to curve in the same direction that the sun curves on the days after the equinox (revisit any of the SunEarth graphs). This theory perhaps also explains why the platform wall itself was reshaped to be so straight (as displayed in Chap. 8).

I do not have any photos showing sun rays overlaying the gouges in the platform wall, and these would be hard to verify for future scholars because the gouges are reburied. I can only suspect that the wall gouges—which themselves show clear evidence of being intentionally made—were perhaps done so to track sun daggers approaching the northern (summer) solstice sunrise. Or perhaps a number of days (49?) before the solstice, similar to the main painted grid. I do plan to return one day, arriving before dawn on days around the equinox and the June solstice to observe and test this theory. I would love to hear from other scholars who get there before I return.

Remaining, however, is the possibility that the triangular window and the sun dagger events are related to days approaching the northern solstice. If the main grid image represents a tally of sunsets intersecting the sun-capture window weeks before the southern solstice, while the *hook head* painting might serve as a symbol of the equinox, it suggests that the inhabitants not only observed the movement of the sun but intentionally altered the outcrop to emphasize those observations. The *lollipop matrix* could represent the motion of sunrises over an entire solar year, positioned in front of the summer solstice sunrise. Note how the sides of the matrix converge as they move upward on the rock, resembling the east side of the *SunEarth* graphs.

This implies that the ancient artist(s) successfully tracked the sun's daily movement across the sky—effectively delineating a solar year similar to that observed at *Serra da Lua/Serra do Sol*—but with a stronger emphasis on sunrises and mornings. Whether the ancients achieved precise alignments or if that precision has eroded over millennia is less important than the overarching evidence. There is sufficient indication to conclude that the Paleoindians here were actively tracking the sun's movements and recording them in rock art. Doing so likely enhanced their predictive understanding of climate cycles and ecological resources.

But why did they do this? Was it an organic development, aimed at improving foraging efficiency, just before the *Younger Dryas* disrupted stable environmental cycles? Or did the inhabitants start tracking the sun because they noticed subtle changes in the environment? While changes in the Amazon may have been less drastic compared to northern latitudes, the timing seems remarkably coincidental. Moreover, evidence of habitation here appears to cease with the onset of the *Younger Dryas*. Did the people perish, migrate, or persist nearby leaving no material record until 6000 years later?

References

Aveni, A., and H. Hartung. 1981. The Observation of the Sun at the Time of Passage through the Zenith in Mesoamerica. *Journal for the History of Astronomy Supplement* 12:S51.

Hugh-Jones, S. 1982. The Pleiades and Scorpius in Barasana Cosmology. In *Ethnoastronomy and Archaeoastronomy in the American Tropics*, ed. A. F. Aveni and G. Urton, 183–201. New York: Annals of the New York Academy of Sciences.

Jansen, M. 1991. Mixtec Pictography: Conventions and Contents. In *Supplement to the Handbook of Middle American Indians, Volume 5: Epigraphy*, 20–33. Austin, TX: University of Texas Press.

Kelley, D., and E. Milone. 2011. *Exploring Ancient Skies: A Survey of Ancient and Cultural Astronomy*. 2nd ed. New York: Springer.

Krupp, E. C. 2012. *Echoes of the Ancient Skies: The Astronomy of Lost Civilizations*. New York: Dover Publications.

Lima, F. P., and S. F. d. M. Figueirôa. 2010. Etnoastronomia No Brasil: A Contribuição De Charles Frederick Hartt E José Vieira Couto De Magalhães. *Boletim do Museu Goeldi (Museu Paraense) de Ciencias Humanas* 5 (2): 295–313.

López-Lanús, B., I. Roesler, D. E. Blanco, P. F. Petracci, M. Serra, and M. E. Zaccagnini. 2007. Bobolink (Dolichonyx Oryzivorus) Numbers and Non-Breeding Ecology in the Rice Fields of San Javier, Santa Fe Province, Argentina. *Ornitologia Neotropical* 18:493–502.

Magalhães, J. V. C. d. 1935 *O Selvagem*. 3rd ed 53. Companhia Ed. São Paulo: Nacional.

NCEI, N. C. f. E. I. 2016. *Magnetic Field Calculators*. Boulder, CO: National Oceanic and Atmospheric Administration (NOAA) Satellite and Information Service (NESDIS).

Schaeffer, D. 2017. Sojourn to the Sun God: Places of Emergence and Movement in Mixtec Codices. Florida State University Libraries Electronics Theses, Treatises and Dissertations. Florida State University Libraries.

SunEarthTools.com 2024 Annual Sun Path. 2024 ed.

Chapter 10
Descendants or Inheritors?

Amazonian Culture and Cosmology

Perhaps the most relevant ethnographic information we have for understanding the ancient *Monte Alegre* artists and their beliefs comes from contact-period and modern ethnographic records of cultures in the Amazon and Orinoco regions. While we cannot determine the extent to which native cultures of the contact period were direct descendants or indirect inheritors of the Paleoindians from 13,000 years ago, the Amazonian culture hearth—despite its diversity—shared numerous cosmological, mythical, linguistic, and material culture attributes. This chapter explores evidence of the earliest *sun/moon worshippers*, either through direct or indirect references to pre-existing (*red* and/or *yellow painted*) rock art or through references to Paleoindian cultures that dwelled in or periodically frequented hilly or mountainous regions.

At this location, current evidence is discontinuous by about 6000 years. Even the contact-period records refer imprecisely to the people of the hills as *Serranos*, which loosely translates to "hill folk." Missionary accounts do not document native explanations for the significance of the art, but there is evidence of European graffiti—including a date of *1764*—among the rock art at *Serra da Lua*. Some of this graffiti appears to have been created using the same paint-processing methods as the ancient art, suggesting that some form of cultural knowledge may have been inherited over time. The missionaries also capitalized on the excellent artistry of local labor and the resilient painting traditions by selling painted, water-resistant gourds.

While no detailed chronicles of missionary work with the *Gurupatuba Indians* at *Monte Alegre* are currently known, there are detailed accounts of the customs of the *Tapajós Indians*, who inhabited the area now known as Santarém over 500 years ago. Earlier observations also exist of *Tupinambá Indian* customs encountered by the Portuguese in the Amazon Delta and as far south as modern-day Rio de Janeiro in *A.D. 1502*.

C. S. Davis, *Before Sunset*, Conflict, Environment, and Social Complexity,
https://doi.org/10.1007/978-3-031-93373-8_10

Nearly all the information in this section comes from culturally and often religiously biased European descriptions and scattered reports, some of which had been collected and retold several years later. In some of these accounts, the chroniclers were describing the traits and rituals of their allies; in other accounts they were describing their enemies. These accounts are not presented here as fact, but rather with the understanding that European chroniclers and missionaries had varied motives for describing cultures and customs foreign to them, and sometimes in direct opposition to their own beliefs. And sometimes while trying to denigrate the native people to devil worship and inhumanity, some accounts possibly did record accurate accounts of what the Europeans saw, even if they misunderstood the meanings. However, for many instances this is the only detail we have of the beliefs and examples of pre-contact South American cultural traditions.

We know from historical records that the first Europeans (*Flemmings*) mentioned in the *Monte Alegre* region traveled there to trade with the *Tapajós* culture (Guedes 1968:56, plate 13; Tyacke 1980:84). Additionally, *Tapajós* (*Santarém*)-style pottery has been identified in the *Monte Alegre* region (Palmatary 1960:19, 90). The *Serranos Indians* were identified as the group that occupied the hills of *Monte Alegre*, and members of this group were taken to the *Tapajós* mission at *Santarém* in the seventeenth century (Palmatary 1960:14). These connections strongly suggest that *Tapajós* beliefs, as chronicled at the mission by Johann Philipp Betendorf after 1661, may share elements with the beliefs of the *Serranos*. This similarity may have developed through proximity and cultural exchange between the two groups even before Europeans arrived on the continent.

Amazonian Linguistic and Archaeological Cultures

Recent linguistic studies on historical place names and the names of various material culture items chronicled in early European accounts indicate that, at the time of contact in the sixteenth century, *Monte Alegre* lay in a narrow corridor of land occupied by Arawakan speakers with a *Mapidian* dialect[1] (Gippert 1993–2010). However, this narrow corridor was bordered by Carib speakers of the *Apalaí* dialect to the northeast, on the same side of the Amazon River. Tupi speakers of the *Nheêngatu* dialect lived in a pocket of land just to the northwest, as well as across the river to the east and south. Tupi speakers of the *Sateré-Mawé* dialect lived across the river to the west and southwest, including the area of *Santarém* (Glück 2000).

Additionally, Gê-speaking Indians occupied much of the interior land further south of the Amazon River, having been pushed inland by Tupi invasions in the centuries before European contact (Freire 2001:9). It is possible that Gê groups occupied the research region decades to centuries before the contact period.

[1] Maps and more details are available from TITUS at <http://titus.uni-frankfurt.de/didact/karten/amer/samerim.htm> Grippert, Jost (c) 1999–2003. Accessed 18 Dec 2013.

Mainland Carib and Arawak tribes inhabited lands further north of *Monte Alegre*, spanning into the Guiana Highlands and the Orinoco River system (Mason 1950; Grimes et al. 1996).

Some linguists who consider Tupi, Carib, and Gê cultures to all stem from a common genealogy (Rodrigues 1985) had proposed a mechanism explaining how such linguistic diversification might have arisen. They describe a proto-language culture to have existed in south-central Brazil, the Guyana uplands, and the east-central Brazilian uplands (Hornborg and Eriksen 2011:146 n.142), all notably containing higher elevation hilly terrain like Monte Alegre. Over time, riverine Arawakan speakers infiltrated and settled along the floodplains of the Orinoco, Negro, and Purús Rivers (Hornborg 2005), eventually circumscribing each of those locations, isolating Caribs in the north, Tupi in the south, and Gê in the east. Arawak language might have become the lingua franca of all these people, while over time each language hearth was left to evolve in isolation (Hornborg and Eriksen 2011:147 n.142).

More recent DNA studies (Raghavan et al. 2015) have given rise to an emerging *Dual Migration Model*. One group of South Americans, more closely associated with Paleoindians, is linked to the *Lagoa Santa* skulls (Moreno-Mayar et al. 2018) and a possible Australasian heritage (Ribeiro-dos-Santos et al. 2020; Skoglund et al. 2015). In contrast, the majority of DNA aligns with Amerindian skulls (Ferraz et al. 2023; Potter et al. 2018; Waters 2019), which many scholars believe represent the only contribution to subsequent native peoples (e.g., *Anzick*) from ancestral Northeast Asian migrations into Beringia—where considerable time led to genetic drift (Llamas et al. 2016). The issue remains unresolved, as some scholars argue that the *Lagoa Santa* skulls and other outliers merely reflect isolation and genetic drift (Moore 2014:81).

The archaeological record for Amazonia indicates the earliest occupations began approximately 13,000 years ago in the lower and central Amazon. Broad-spectrum tropical forest foragers used stemmed bifacial projectile points (Costa 2009; Góes and Petersen 2006; Roosevelt et al. 1996). Recent discoveries have extended the time frame of Paleoindian presence at some Brazilian sites to 15,500–12,800 calibrated years BP (Bueno and Isnardis 2018; Prates et al. 2020). Most upper Amazon sites appear slightly later, around 11,000 years ago (Barse 1990; Magalhães 1994; Meggers and Miller 2003; Prous and Fogaça 1999). However, research at *Serranía La Lindosa* has drawn attention to rock art with artistic similarities to *Monte Alegre*, dated to approximately the same age—13,000 years ago (Morcote-Ríos et al. 2021).

Shell mounds created by marine and riverine foragers begin to appear along the Atlantic coast and the Lower Amazon around 9500 years ago, accompanied by the earliest ceramics between 6000 and 8000 years ago (Bandeira 2008; Gaspar et al. 2008; Roosevelt 1991; Rostain 2008b). Around the same time, horticulture emerged in the upper Madeira River region (Miller 1999). Approximately 2500 years ago, root crop domestication and slash-and-burn agriculture began in small garden plots, eventually expanding to semi-intensive communal fields across numerous Amazonian cultures during the Late Holocene (Denevan 2001; Oliver 2008). This expansion was coupled with deliberate tree planting and wild animal management

(Clement 1999, 2006). Additionally, some regions developed sophisticated fish-farming techniques (Erickson 2000), while the savanna wetlands of southwestern and northeastern Amazonia adopted raised-mound agricultural techniques (Erickson 2008; Rostain 2008a).

Marajó Island, at the mouth of the Amazon River, was inhabited by Arawakan and Tupi speakers at the time of European contact. However, the archaeological record of the island shows a long chronology based on ceramic assemblages beginning with Ananatuba (~3500 to ~3000 BP), Mangueiras (~3000 to ~2800 BP), Formiga (~AD 0 to 800), Marajoara phase (~AD 400 to ~1300), and Aruã (~AD 1300 to ~1600) (Meggers and Danon 1988; Roosevelt 1991; Roosevelt 1993; Schaan 2001, 2004; Simões 1969). The interior of the island contained the largest known archaeological mound on the island (Camutins site) with Marajoara ceramics and associated burial urn and cemetery practices commencing about AD 400–600 (Schaan 2004).

Similar necropolises have been found on the coast of the Guianas, and on the coast of Amapá on the Brazilian side of the border (Guapindaia 2001). However, the origin of burying the dead directly in burial urns has been attributed to the Guarani (Boglar 1958; Chaumeil 2007:251; Nordenskiöld 1920). Regardless, the "chiefdom-level" Marajoara pottery style at the mouth of the Amazon, with its modeled, incised, and bi-chrome decoration, might represent an ethnogenesis, combining elements of Amazonian Barrancoid pottery influence (first emerging 2500 years ago) with southeast Amazonian Tupi-Guarani painted pottery.

Regional variants of this new Amazonian Polychrome tradition diffused far upriver by about AD 1000–1250, influencing the later Santarém-style pottery used by the *Tapajós* culture at the time of European contact, which also shows significant earlier influence from (possibly Carib-Speaking) Arauquinoid complex pottery styles from the Orinoco and Guianas regions (Heckenberger and Neves 2009:255–256). Similarly, recent evidence of cultural contact, based on Incised Punctate style blended with Arauquinoid pottery styles at the Maranhão stilt village site of Formoso, suggests ceremonial trade and ritual transmission of astronomy knowledge may have taken place as early as the fifth century A.D. (Davis and Navarro 2023; Navarro 2022).

By examining Tupi, *Tapajos*, Gê, and Highland Guiana cosmology, we can hone in on regional beliefs that were more likely shared among the *Serranos* that inhabited the research area. However, precautions remain that all these contact-period cultures are separated from the earliest rock art painters of Monte Alegre by more than 10,000 years.

Only the most rudimentary beliefs might have passed down, if at all. It is more likely that these recent cultures modified their beliefs in response to each of their specific culture- histories, knowledge of their surroundings, and perhaps ever-so-subtle influence by ancient material evidence left by long departed Paleoindian inhabitants.

The Tupinambá and Tupi-related Groups

While the *Tupinambá*[2] were a specific tribe far from the research area, European contact with them contains some of the earliest descriptions of South American customs before European contact took on widespread influence. The Indians occupying the Amazon Delta in the sixteenth century were identified as the *Tupinambá* historically and archaeologically by their pottery with distinctive polychrome decoration, which completely replaced the previous shell-mound builders just before the beginning of the Christian era (Gaspar et al. 2008; Neves 2011:37; Noelli 2008).

The *Tupinambá* spoke the same Tupi language as many tribes living along the Atlantic coast and numerous riverbanks and river islands in the lower Amazon Basin. The native language of the *Tapajós* culture at the time of earliest contact remains unknown, but Curt Nimuendajú (1949:95) argued that they did not speak Tupi and noted that several place names appeared to be of Carib origin. However, Frederico Barata (1950:464) contended that the extensive trade network associated with *Tapajós* pottery implies frequent contact with Tupi-speaking groups and suggests that the *Tapajós* may have spoken Tupi themselves (Palmatary 1960:14).

Herbert Smith summarized regional myths recorded by previous explorers, including the Tupi belief that all things were created by either the sun, the moon, or a third water deity (Smith 1879:586, 541–587). In the sixteenth century, Jean de Léry added that, "not only do [the *Tupinambá*] believe in the immortality of souls, but they also firmly maintain that after the death of bodies, the souls of those who have lived virtuously go off behind the high mountains where they dance in beautiful gardens with the souls of their forbears. On the contrary, the souls of the effeminate and worthless, who have neglected the defense of their fatherland, go with Aygnan, by whom, they say, these unworthy ones are incessantly tormented" (Léry 1992:91, 136).

For Tupi-speaking tribes, Tupa is their word for father, for the Supreme Being, and for thunder. This was no coincidence as they viewed thunder (and other natural inconveniences) as the representation of disapproval by their deity, in the same way that an elder might disapprove of a child's behavior. Yet, despite their belief in a single "Universal Parent," they addressed no prayers to it, showing neither fear nor placing hope in it (although they feared thunder). Instead, superstitions dominated their faith, which their *Payes*[3] were said to be able to interpret from dreams, shadows, nightmares, hallucinations, and especially from the maracas (Southey

[2] The *Tupinambá* have a unique colonial history that might be more legend than fact. Supposedly after defeat and banishment from their homeland near modern-day Rio de Janeiro in the early sixteenth century, a group made their way north via the Andes mountains and they settled near the Spanish controlled town of Quito for some time, until one of their members stirred the ire of the Spaniards by stealing livestock. The tribe then fled down the Amazon River and settled on the Island of *Tupinambá* in the middle of the river near the confluence of the Amazon with the Madeira River.

[3] *Paye* is the Tupi word for their priests, who the missionaries regarded negatively as "quacks, jugglers," and charlatans (Southey 1801:227).

1810:227). It was the duty of the *Payes* to ward off evil spirits, which the Tupi believed were constantly lurking in the dark. For this reason, "they kept fire burning all through the night in their houses, for which their chief reason was that Evil Spirits could not come near it; nor would [the Tupi], if it could be avoided, ever venture out in darkness without a firebrand" (Southey 1810:228).

The *Tupinambá* also made predictions based on superstitious occurrences, most of which were interpreted as curses or signs of bad luck. For example, they believed the moon was female, and her disfavor could bring sickness and misfortune (Southey 1810:193, 195, 228). They took great care to avoid killing female animals or eating eggs while their wives were pregnant, fearing karmic retribution affecting their own children (Southey 1810:239).

Tupinambá women were the sole artisans of spinning and weaving. Using a spindle whorl, they spun cotton even as they walked. The cotton string was woven into sturdy hammocks, which the women later cleaned by cutting and boiling a gourd to produce a lather that made the cotton as white as snow. These women were also skillful potters, drying their vessels in the sun, inverting them, and firing them under a pile of ignited bark. Some tribes created pots large enough to bury their dead, placing the bodies upright within the jars. The *Tupinambá* women were additionally known for coating the insides of their cooking vessels with a white glaze, possibly made from lime, and for crafting wicker and straw baskets. Men, however, were primarily engaged in war, hunting, and chopping down trees to make canoes. They also fished by shooting long arrows into the water (Southey 1810:243–244).

Their manner of burial was also rooted in superstition: a round pit grave was dug by the nearest relative and wooden stakes were erected as posts to form a vault; a hammock was then suspended from the posts and the body was placed sitting within it; food was placed before it, and if it was male, so too his maraca and bow (Southey 1810:247–248). The limbs of the deceased body were tied so that the corpse would not be able to rise later and infest friends and family that it wished to visit; it was believed that the spirit would wander, particularly to the mountains for sport, but return to its grave to eat and rest (Southey 1810:247–248). Also, it was their belief that a nocturnal bird no bigger than a pigeon and plumed with ash-gray feathers, and whose chirp was more penetrating even than the screech owl, was sent by their dead kinsmen as a sign of good luck and encouragement in battle; and so they neither harmed the bird, nor let knowledge of others having harmed the bird go without making them repent (Léry 1992:91).

The *Tupinambá* had a culture heavily steeped in animistic beliefs. The most significant information with relevance to the rock art of Monte Alegre is their belief that the spirits of their dead ancestors join with one another behind the mountains to dance and rejoice. Considering this, hills and mountains were probably avoided by most, perhaps except for the holy men. But more importantly, we should consider how this idea came about. Is there something empirical about mountains that led them to believe their deceased would go there? Except for possible associations between death and caves on a mountain, the *Tupinambá* belief is hardly universal. Perhaps their beliefs came from them discovering Paleoindian rock art left by people long gone that they had little knowledge of.

The Tapajos *Indians*

Although the *Tapajós Indians* might have known and spoken the Tupi language, they inhabited lands near the confluence of the Tapajós and Amazon Rivers. At the time of historic contact, this area appears to have been devoid of the *Tupinambá* polychrome-style pottery (Neves 2011:46). Archaeologically, *Tapajós* pottery is associated with an incised-punctated style (Kondurí and Tapajó), dating back to the end of the first millennium A.D. (Gomes 2002:131). This style shares the most similarity with the AD 400 to AD 1400 *Arauquinoid* series pottery of the middle Orinoco (Navarrete 1999:41; Zucchi 1985), coastal Suriname, and coastal French Guyana (Rostain 1994:84; Rostain and Versteeg 2004:239). Assuming the pottery style originated from tribally related people, the *Santarém* area may have been its greatest concentration south of the Guyana Plateau, suggesting a radial relation or diffusion of Carib groups from that region through *Monte Alegre* and down to the *Tapajós Indians* occupying the *Santarém* area at the time of European contact.

Unfortunately, by the time chroniclers were living among the *Tapajós Indians* and recording their customs and beliefs, the community had already been drastically affected by European arrival. Survivors of war, disease, and enslavement were herded into missions where they were subjected to Christian conversion and cultural suppression. Nonetheless, much of the following information comes from missionaries who caught their subjects covertly practicing traditional customs. While these accounts should be interpreted with caution due to the varying cultural contexts and historical settings, significant parallels in cosmological views suggest common themes that may be useful in interpreting the *Monte Alegre* rock art.

The Jesuit missionary Johann Philipp Bettendorf (1909:154) described ancestor worship among the *Tapajós Indians* during his stay in 1661. He noted that they had a dried body, called *Monhangarypy*,[4] suspended in a box for many years under the roof of one of their houses. They worshipped this "ancestor" with gifts and dances. Ancestor worship may have been widespread and long-enduring in the Amazon basin, as subsequent missionaries repeatedly discovered that this practice had been revived, even after Friar Antonio Pereira ordered the house burned to the ground.

Around 1700, Friar Manuel Rebello reported that the *Tapajós* had dried bodies that they worshipped as idols. Each represented different aspects of life: one as the god of maize, another as the god of manioc, another for the sun, the rain, and so on (Palmatary 1960:15). In 1742, Friar Luiz Álvarez confiscated 16 dead bodies from the *Tapajós* and *Caianas* tribes. The Indians claimed these were the bodies of the first peoples of the world (Palmatary 1960:16).

Friar João Daniel recorded enduring ancestor worship during his missionary work in the Amazon from A.D. 1751 to 1757—long after much of the indigenous population had died or been replaced by enslaved Indians from various tribes throughout Brazil. At an unspecified mission, Friar Daniel noted that the Indians

---

[4] In their language this meant "first father", but the custom sounds incredibly Arawakan, as very similar customs were recorded on Hispaniola by Friar Ramon Pané (1999).

kept seven mummies in a secret hut in the forest, known only to the elders. Once a year, they slipped away to re-sew the mummies into new clothes. In the same house were five stones, which were equally venerated for specific purposes: one for weddings, another for childbirth, and so on (Nimuendajú 1952:7).

According to Friar João Daniel (1841:168), the *Arapiuns* Indians,[5] who also lived on the Tapajos River for many years, shared the custom of endocannibalism with the *Tapajos*— by desiccating their dead in a separate hut and then pulverizing the bones to mix them into their fermented maize that the elders drank during festivals. However, Palmatary cites Daniel also describing how this tribe:

> … worshipped the stars, the sun, and the moon, but mostly the sun and the moon. The appearance of the new moon was celebrated by a festival. Their name for the sun was Coara Cy, mother of the day, mother of the earth. Their name for the moon was Ja Cy, mother of the fruits of the earth. These two heavenly bodies were regarded as divinities—creators. The missionary once saw an Indian come out of a large hut and shout that he saw the new moon. Immediately all the other occupants rushed out, stretched their bodies, arms, and hands upwards as if praying for health and strength. Other expressions of their adoration of the moon were jumping for joy, shouts of welcome and presenting to it their children. [Palmatary 1960:16]

For themselves, the *Tapajos* Indians wove constrictive arm, leg, and headbands, and they wore ear ornaments (Palmatary 1960:17). However, none of these items were more prestigious or mysteries than the *muiraquitãs*[6]—greenstone amulets believed to be made of jade, nephrite, or serpentine, and usually fashioned to resemble batrachian forms. These talismans carry important ethnographic significance due to their possible connection to a sun- and/or moon-worshipping matriarchal society.

From *Cônego* Francisco Bernardino de Souza's *Lembranças e Curiosidades de Valle do Amazonas* published in 1873, we have a sense of how widespread the South American legend was. The true *muiraquitã* amulets of legend were formed from clay at the bottom of a lake *Yaci-uaruá* ("mirror of the moon") near the headwaters of the Nhamundá River during a full moon, at the time of year when a certain frog species chirps its mating call (presumably March–May). The *Icamiabas*[7] (women without husbands—i.e., the Amazons[8]) celebrated a festival to the moon and to the mother of the *muiraquitã*, who lives in the lake. The clay was shaped underwater and hardened when exposed to air. Despite the legendary creators of these stones

[5] The Arapiuns River is a black water tributary of the Tapajos River, and their confluence meets about 30 kilometers before the Tapajos combines with the Amazon.

[6] Early spellings include *buiraquitas*.

[7] According to the naturalist João Barbosa Rodrigues (1875:33), *Icamiaba* is a shortening of a Tupi word—*Itacamiaba*—which means "gift of stone from the woodland," but anthropologist Federico Barata (1954) believes "Itã" means "toad" or "frog," referencing the *Tembé* Indians of Gurupi.

[8] Rodrigues (1875:7,31,53) claimed to have found the village of the Amazons 18 miles from the mouth of the Trombetas River, near a lake called "Costa do Parú," but the moon lake was surrounded by forest-covered mountains in the Nhamundá valley. He also claimed that the Amazon tribe taught their lapidary skills to Indians of the Uaupés River and that the two tribes might have merged by the time of Rodrigues' exploration.

being the *Amazons*, most historical references locate trade for them as radiating from the *Tapajos* Indians (Palmatary 1960:76–78).

The legend of the *Amazon* women and their *muiraquitãs* might have spread as far as Hispaniola in the Caribbean, with the story of the women of *Matinino* mimicking that of the *Icamiabas*. This bears linguistic importance because Arawakan speakers (the *Taino*) inhabited Hispaniola at the time of European contact, and they share strikingly similar idol/ancestor worship to the *Tapajos* Indians in the myths recorded and in the religious idols (called *zemis)* confiscated by the earliest missionaries there (Pané 1999). With the *Tapajos* tribe living so close, Arawakan culture might have inhabited the region near Monte Alegre at this time as well.

In *British Guiana*, chroniclers recorded most stories about Amazon women from the *Arawak* and *Macusi Indians*. *Muiraquitãs*—jade-like amulets—have been found throughout Brazil, particularly along major rivers, and even among the *Caribs* along the Guiana coast, where they were called *macuaba* or *calicot stones*. On the *Orinoco River*, they were referred to as *macagua*, a variant of the Guiana term. Europeans also sought to trade for these stones, and eastern Brazilian tribes reportedly used them as dowries to acquire wives. Although natural jadeite is rarely found in Brazil, a large block of nephrite weighing almost six kilograms was reported to have arrived in *Manaus* from the *Roraima* mountain range in 1928 (Palmatary 1960:80).

The stones not only represent material evidence of the legendary Amazonian tribes but also signify the widespread prominence of women in the societal frameworks of many indigenous South American cultures. Orellana's chronicle of his journey has faced significant criticism, with claims that it was exaggerated to absolve him of desertion charges by *Pizarro's* camp. Much of the skepticism centers on the perceived incredulity of his tale about Amazonian women warriors. However, in his account, Orellana described interrogating an Indian trumpeter girl using a prepared vocabulary. From the prisoner, he claimed to have learned that:

> "This country was subject to women led by a chief called Caripuna, who lived after the manner of the Amazons of the [ancient Greek legend], and who possessed gold and silver in abundance. There were in their dominions five temples of the sun, all covered with plates of gold; their houses were of stone, and their cities walled. They lived on the western side of a large lake, which they called the Mansion of the Sun, because the sun sunk into it" (Southey 1810:96, 159).

The information presented here further suggests that the *Tapajós Indians* were likely an Arawak-speaking culture who frequently interacted with Tupi-speaking groups.

Religious elements among the *Tapajós* and *Tupinambá* both emphasize ancestor worship as the central aspect of their animistic belief systems. However, neither group fits the description of an alternative sun/moon-worshiping culture, nor do they provide substantial insight into the rock art symbols at *Monte Alegre*. There is, however, another language group that some believe inhabited much of the region before the Tupi displaced them. This group, referred to as the *macro-Gê*, preserves myths that may offer insights into the connections between the sun, the moon, ancestor worship, and the acknowledgment of the first peoples.

Gê Cultures

The Gê Indians are comprised of several groups of South American Indians currently found south of the Amazon River and generally east of the Xingu River, although this group is often further divided between the *Cayapo* groups of northern Gê speakers in central Brazil and the *Timbira* groups in the eastern high steppes of Brazil nearer to the Atlantic Ocean.

Each of those geographical groups is further subdivided into numerous tribes. Yet, despite the diversity of tribes, Gê groups share more than just a common language family. Their myths also share a "remarkably high degree of consistency in both form and content" (Levi-Strauss 1969; Wilbert and Simoneau 1978:22). Among the Gê-speaking Indians living in the interior east central of Brazil, several ethnographers in the early twentieth century recorded creation myths referring to a time when the sun and moon were people living on the surface of the earth (Roth 1915:254; Villas Boas and Villas Boas 1970:57–97; Wilbert and Simoneau 1987:31–106). Despite detail variations among cultures, the myths commonly describe the Sun acquiring a red crown from a plumed woodpecker (Wilbert and Simoneau 1987:33, 41, 52, 59, 63). These myths typically continue to explain that the Sun acquires red feces by eating ripe fruit from a *Buriti* palm (*Mauritia flexuosa*-also called *Moriche or Miriti)*, a fruit with a red covering and yellow interior.

Populations living along the Amazon River still make a drink from this fruit today, and they believe it is good for soothing burns. Around the village of *Paituna* the palm tree begins to bear fruit in November/December (Cavalcante 1988:168–171), a time frame coinciding with 49 days before the southern solstice.

Just as Alfred Russell Wallace (1889:103–104) reported a group of Buriti palms in the hills of Monte Alegre while traveling between rock art sites, I also found a single Buriti palm marking a natural water spring trickling out the hillside a few hundred meters south of the *Pedra da Tartaruga* rock art site. Buriti palm seeds favor growth in water and are often only found in pre-ceramic phase archaeological contexts (Oliver 2008:199). Although the tree is recent, its lone presence far away from the swamp is most likely due to deliberate seeding to mark the location of the spring water. Similar sources of drinkable water were, no doubt, available in the hills in the past—their locations unavoidably marked by Buriti palms, an oligarch palm species (Peters 2000). So well-known is the palm tree's association with fresh water that Brazilians today refer to it as the "tree of life" (Erickson 2008:176).

In the same Gê creation myths of the Sun, his companion, the Moon, always impulsively mimics him. The Moon constantly attempts to acquire the same qualities as the Sun, but he finds the woodpecker feathers too hot to handle (causing a brush fire). He similarly finds the palm tree fruit to be unripe (Wilbert and Simoneau 1987:34, 44, 53, 60). Nimuendajú (1939:158–165) even recorded an *Apinayé*, *Serente*, and *Ramkokamekra* version of these myths that accredits the Sun and Moon with creating people, and for this reason, the villages of those cultures are divided into sun and moon moieties, with the sun group forever leading the village (Wilbert and Simoneau 1987:56).

A few other stories about the sun and the moon appear in *Gê* myths. Some of these myths attempt to explain the difference in intensity and size of the sun throughout the day (Reichel-Dolmatoff 1996:32). From the *Suya* example, we see how daily activities are structured by the sun's position in the sky—a relationship likely shared by nearly all South American cultures. Interestingly, myths about the sun and moon are typically among the earliest cosmological narratives (Civrieux 1980:19–32; Reichel-Dolmatoff 1971:24; Roth 1915:254; Wilbert and Simoneau 1987:32–73). While the origins of these myths remain unknown, they often describe a time before the existence of people, when the Sun and Moon were hunters roaming the world in search of food and other adventures (Villas Boas and Villas Boas 1970:57–93).

From the *Gê* cultures (notably the *Kayapó*), we gain insight into the ancient world in which these myths evolved. The reference to the Sun and Moon as the first people in the world, leaving red stains across the land, could be interpreted as an allusion to the red and yellow lichen that "stains" trees and rocks. Alternatively, it might reference rock art left by an earlier culture—art that the *Gê* perhaps viewed as evidence of the Sun and Moon's earthly presence long before their time. If we assume these myths contain even a small reference to physical reality, it is plausible that pre-existing sites like *Monte Alegre* inspired aspects of *Gê* cosmology.

Some myths shared by tribes along the *Xingu River* recount the Sun's death, plunging the world into darkness until his three sons took over their father's duties. None of the sons could manage the task alone, so they divided the responsibilities, each enduring the heat and light of the sun at different times of the day or under varying intensities (Villas Boas and Villas Boas 1970:94–97). Similar myths appear in the traditions of several cultures across North and South America, including the *Maya*, *Aztec*, *Navajo*, *Iroquois*, and others. Many Amazonian tribes also recount a period of prolonged darkness and associate eclipses with battles between the Sun and Moon (Smith 1879:392; Wilbert and Simoneau 1987:94). These beliefs persist among *Gê* and Arawak-speaking cultures, particularly in remote forest pockets that remained virtually uncontacted by Western civilization until the twentieth century.

Another common theme in *Gê* mythology describes a time when people did not cook their food but instead left meat in the sun to be heated (Wilbert and Simoneau 1978). In the early 1970s, ethnomusicologist Anthony Seeger (1981) lived with a northern *Gê* group he referred to as the *Suya*, named for their oral songs used to pass down tribal knowledge. Although this account is more recent, it remains one of the few detailed ethnographies of a group that avoided contact with Western civilization until the late twentieth century. *Suya* myths accurately outline daily routines and gender roles, consistent with Seeger's direct observations.

A *Suya* day begins with all the people bathing in the river at dawn; then, men hunt or fish in the morning while women prepare the first meal. All take naps afterward, during the heat of the day. Later, families play with children in the afternoon, and men gather to talk or sing at dusk. In the early evening older men go off to bed while young men stay awake courting women (Seeger 1981:13–14). When the men would sing at dusk women would join in with them; their songs, as well as their minds—Seeger claims—were one-tracked or contextual. If a particular event occurred, they would recall songs only related to information about it, seeming unwilling or even incapable of changing subjects (Seeger 1981:15, 164). Anyone who would speak about negative things like witchcraft or other taboo subjects would invoke a sense of dread among the village for fear of repercussions (Seeger 1981:17).

The *Suya* did not have the same concept of heaven and hell or good and bad that Western cultures had, nor did they show signs of worshipping gods. Instead, moral concepts were related to social/pleasant versus antisocial/unpleasant human faculties. Good moral behavior was equated with cleanliness and friendliness, but immoral behavior was equated with malodorous and vitriolic behavior (Seeger 1981:27–28). Animals were similarly viewed in terms of their odor; those with strong odors were considered powerful and dangerous to humans, a belief system they applied even to the cooked meat of animals. This they claimed to have learned from the jaguar (Seeger 1981:23). It is important to note, however, that in Gê myths animals are often used as metaphors in important rituals and in origin stories.

Stephen Fabian (2001:75–76) reports that the *Bororo*—*Gê*-speaking Indians of Brazil's *Mato Grosso*—used anaconda fat as the preferred medium for mixing and applying red dye as body paint during ritual contexts. *Jure*, the *Bororo* word for the great anaconda snake, holds deep cultural significance, symbolizing ritual, dance, beauty, fecundity, vitality, rainbows, rain, and rivers. Because anacondas are often found in or near rivers, the *Bororo* associate the snake with rain, particularly the rainy season, which floods the rivers, creates rainbows, and revitalizes the land with fertile plants and breeding animals.

Fabian explains:

> "*Jure* is also a serpentine dance for maturing youths and adults. The dance is associated by myth to the maturation of a young male *Bororo* culture hero… [who] brings ritually significant rattles—the 'noise of culture'—back to his people" (Fabian 2001:76).

This single animal symbol encapsulates the *Gê* tradition of overlaying multiple metaphors between myth, ceremony, meteorology, and environment. The anaconda's significance, along with that of other animals, extends to the naming of certain stellar asterisms by the *Bororo*. This practice connects environmental phenomena (such as the rainy flood season) with cultural ceremonies (such as the rite-of-passage dance) and art (using snake fat as a binder for red paint), all tied to a specific time of the year determined through stellar observation.

Arawak Influence

Many Arawakan language groups are located today in the Southern Amazon at the headwaters of the Tapajos River, the Guaporé River bordering Bolivia, and the upper Paraguay River. Many of these groups—including the *Baúre*, the *Paresí*, and the *Terena/Guana*—have apparently been there at least since the beginning of the seventeenth century when the earliest historical accounts describe the region.

Despite their diversity, Arawak groups have long been noted (Denevan 1966; Métraux 1942; Oberg 1949; Oliveira 1968; Schmidt 1914) for consistency of shared social organization, village settlement patterns, rituals involving sacred male flutes, and subsistence technology, which includes sophisticated community fish weirs and traps, dunk baskets, and river poisonings (Erickson 2000; Heckenberger 2011:61). Basketry and hammock-making styles, when available, also correlate well with language-group identifications. Cordage spin/twist patterns relate to basic motor habits, with a right-side downward slant S-spin/twist style associated with Arawak groups, and a left-side downward slant Z-spin/twist style associated with Tupi, Carib, and Gê groups (Heckenberger 2011:62; Petersen et al. 2001).

Arawakan village patterns provide significant insight into their cosmology. They are arranged in circular plaza villages with central ceremonial structures, notably representative of a concentric circle or dot if viewed from above. According to Michael Heckenberger (2011:62), "this appears to be a feature of early Arawak proto-culture, as known from earliest plaza villages associated with Saladoid in the Caribbean (Trants site in Montserrat) and Amazonian Barrancoid in the central Amazon (Osvaldo site at the confluence of the Negro and Solimões Rivers in Brazil) and upper Amazon (*Hupa-iya* site on the Ucayali River in Peru), 500–200 B.C." These plaza villages are also associated with a "temple-idol-priest" ritual complex where—unlike the central structure of other Amazonian villages being the men's house—the central temple serves as a repository for sacred ritual masks, idols, and flutes (Steward and Faron 1959).

While the *Taino* share the temple-idol-priest complex of mainland Arawaks, their creation story connects their ancestors to caves and the sun. The following account was told to Friar Ramon Pané to explain how the *Taino* came to be:

> Hispaniola has a section called Caonao in which there is a mountain called Cauta and it has two caves, Cacibajagua [Cave of the Jagua] and Amayauna [without importance]. From Cacibajagua came most of the people who inhabit the island. While still in the caves, this people kept watch at night and they put in charge someone called Macocael [he of the eyes which do not blink] who was carried off by the sun because he returned to the door late one day. Since the sun had carried him off on account of his poor vigilance, they shut the door. Thus he was turned into stone near the door. The reason that Macocael stayed awake and kept watch was to see whence he would send away and divide the people and it seems his delay was a great mistake. Afterwards others went out to fish and they were made prisoners by the sun, and they were turned into jobos [cherry plum trees]. It came to pass that one man named Guahayona [our pride] said to another named Cahubaba [the ancient one] that they should go harvest digo [a root which they bathe with to wash the body]. Cahubaba went out before daybreak and the sun overtook him on the road and he was turned into the bird that sings in the morning and he is called Cahubabael [son of the ancient one]. When he saw that the one sent out to gather digo did not return, Guahayona decided to leave the cave called Cacibajagua. [Stevens Arroyo 1988]

Jagua is a vegetable used as a black dye by the *Tainos* to paint their bodies. In *Taino* mythology, two caves represent their origins: one cave symbolizes the origin of the *Taino* people, while the other, described as insignificant, symbolizes the origin of all other people. The *Taino* believed their ancestors once lived in the same mountain cave where their body paint originates—a cave where they feared the Sun could capture them. In a similar way that the jaguar in *Gê* myths might symbolize a different culture or a specialized class, the Sun in this myth could carry a similar metaphorical significance. Both archaeologically and metaphorically, Caribbean ancestors appear connected to a mainland origin, and it seems plausible that this myth references an earlier sun-worshipping culture that also lived among the mountains.

Ancestor worship and the housing of religious idols were common traditions among Arawakan cultures, regardless of the distance separating their communities in space or time. Antônio Pires de Campos (1862:443–444) provides a historical account from AD 1720,[9] noting that:

> "[The *Paresí Indians*] idols have a separate house with many figures of varied forms, in which only men are allowed to enter… [They also] make objects of stone like jasper in the form of the Malta cross, an insignia only used by chiefs."

Hereditary priests and nobles were housed in other prominent buildings, while chiefs were further adorned with sun diadems made from harpy eagle feathers dyed yellow (Heckenberger 2011:62).

Southeastern Arawakan cultures, such as those in the upper *Xingu River* region, continue to retain many Arawakan cultural traditions despite living adjacent to *Gê*-speaking cultures. Linguistically, *Xingu Arawaks* appear to have derived from the southwestern branch of the Arawakan language family in ancient times. They can be further divided into a southern branch (*Mehináku/Waurá*) and a northern branch (*Yawalapiti*). Archaeological evidence of their presence in the region dates to just before AD 500, as seen in their agricultural techniques (Heckenberger 2011:63).

One of the most important ceremonies of the *Yawalapiti* tribe is the *Pequi* festival.

Pequi is a tree (*Caryocar brasiliense* or *Caryocar coriaceum*) that bears a bright yellow fruit that is made into a *mingau* (porridge) (Currington 2005). Like the *Jure* ceremony of the (Gê- speaking) *Bororo*, the *Pequi* festival is embedded with rich cultural significance. The *pequi* fruit is used domestically for food, for producing cooking oil or as a substrate, and for folk medicine[10] (Júnior et al. 2013:230). The tree begins to blossom in December and it bears fruit that is harvested in April or May. The *Pequi* festival is held annually at the end of the harvest season. The entire family takes part in harvesting the fruit, and so the entire community also participates in the festival (Gonçalves 2008:91).

The ceremony serves as a rite of passage for girls entering puberty and males seeking wives, as well as a harvest celebration; ultimately, it is a celebration of life. Body paintings are displayed by nearly everyone, with clear distinctions based on gender. Women almost exclusively paint geometric designs, which, while specific to

---

[9] Belonging to the Southern branch of Arawak speakers, more specifically the southwest branch. The *Paresí* have lived in the southern part of the present Mato Grosso state of Brazil, near the border of Bolivia for as long as there is a historic record of them.

[10] It is rich in vitamins A and E, and its leaves are used therapeutically to regulate menstrual flow. The oil is used to treat conditions of the scalp, muscular and rheumatic pain, broncho-pulmonary infections, asthma, and inflammation in animals. The wood produces a yellow varnish and is used in woodwork; the pericarp is rich in tannins used to dye cotton (Gonçalves 2008:91).

the *Xingu* region, have no particular meaning beyond the creativity of the painter (Currington 2005:34.59). Men, on the other hand, typically paint their own bodies with bold applications of red pigment, often in the form of smears, streaks, or handprints (Currington 2005:46.30).

The ceremony involves a variety of activities, including religious rituals, singing, dancing, the use of sacred musical instruments, sports competitions, and "gender wars" (McCallum 2001:160). Uniquely, prizes are awarded to those who harvest both the least and the most amounts (Júnior et al. 2013:230). The mention of harvest competitions links this important festival to seasonal ecology and inevitably yearly repetition, so time reckoning was monitored and collectively declared.

Men also adorn themselves with vibrant yellow headdresses made from the feathers of the toucan bird, from which the name *Yawalapiti* is derived (Currington 2005:37.16). Bird carvings are an exclusively male activity. These carvings often depict hummingbirds (*colibris*), considered their guardians. The wooden birds are always painted red using a wild fruit called *urucum*[11] (Currington 2005:43.04–26).

Pequi fruit is also used as rattles by the *Kamayurá* tribe—Tupi speakers—who tie the dried fruit around the right ankle of dancers during a similar rite-of-passage ceremony (Bastos 2011:77). The *Kamayurá* myth of the origin of the *pequi* fruit is blatantly associated with sexual symbolism, as the tree is said to have sprouted from the ashes of a beautiful young male *jacaré* (crocodile) who had an affair with two married women of the same husband. The husband kills the *jacaré* with an arrow and abandons the women, who cremate the body, causing a *pequi* tree (in a manner that metaphors a cosmic tree or tree of life) to sprout from his ashes and bear fruit with no odor. This fruit draws the attention of a *papagaio* (parrot) whose owner *Murenayat* (a Tupi-speaking landlord of *Murena*—the center of the world) comes and is wooed by the two women because they place the fruit between their thighs. The result of their actions permanently transfers their sexual feminine odor to the *pequi* fruit. *Murenayat* passes this *pequi* fruit to all his lovers so that his infidelity will go undetected. The *Kamayurá* hear of this fruit and they take *pequi* seeds from *Murena* to plant all around their village (Agostinho 2009:87–89).

The *Kamayurá* myth of the *pequi* fruit carries much more symbolism than is apparent from a direct read. There are other characters, such as a spider, that catches *Murenayat*, and there are other symbolic events, such as a line that *Murenayat* attaches to his parrot to fly to the *pequi* tree. These references might be stellar constellation references relating important landscape details to months of the year. The reference to the youthful *jacaré* is pluralistically symbolic as well, indicative of the flooding river, and possibly of a foreign tribe of river- dwellers. The yield of the *pequi* harvest in April/May depends on the abundance of rainfall during the rainy

[11] *Urucum* derives from the Tupi word for the color red– *uru ku*, which is produced from the seeds in the seedpod of the *Bixa orellana* shrub. *Urucum* is better known to North Americans as achiote or annatto, the natural red spice used to color foods and soups. This grows in Monte Alegre and the matron of my host family performed the process for producing the red powder. The process produces an unpleasant but not overwhelming odor during part of the four to five days it demands, two or three of which involve drying the seeds in the sun.

summer season of the preceding year, which raises the levels of the rivers until December (Gonçalves 2008:91).

The profound knowledge that the indigenous community has of the cycles of nature is obvious in their relations with the forest. Such cycles call upon more elaborate strategies of natural resource use and management that the community has accumulated from past generations. This knowledge is spread by careful observations and oral accounts of the dynamic forest environment. In this case, the environment becomes understood through appropriation of nature by the people (Gonçalves 2008:93–94).

Like the *Pequi* festival, the *Yurupary* cult is a male initiation ritual shared by several Tukanoan and Arawakan language cultures in northwestern lowland Amazonia. Stephen Hugh-Jones (1979) attempted an ethnographic report of the cult ritual in its fullest details and explanations, after he witnessed the ceremony practiced by the *Barasana*—a Tukanoan- speaking community living in the Vaupés region of Colombia.

Yurupary[12] is a lingua franca word originating from the Tupian Lingua Geral (Nheêngatu). The word can be used to refer to the exclusive male cult, or as a blanket name for several mythical characters, including a culture hero who is burned alive for indulging in cannibalism only to have his ashes give rise to a "walking" palm tree.[13] Alternatively, *Yurupary* can refer to the sacred flute and trumpet musical instruments made from said palm tree for said ritual (Hugh-Jones 1979:5–7).

Most Tukanoan cultures display gender differences when applying body art for the ceremony. Tukanoan men are accustomed to paint their faces and bodies and make drawings in the sand or on a dusty floor (Reichel-Dolmatoff 1978a). Women will paint themselves or each other and help to paint their husbands (Reichel-Dolmatoff 1978a:21). Little wooden stools are the exclusive property of men and are almost always painted. The stools are used for rest and relaxation but also meditation (Reichel-Dolmatoff 1978b:23).

The *Barasana* also have a metaphorically rich myth about their ancestors. In their story, their earliest forefathers stem from the anaconda ancestral spirit—who, according to Hugh-Jones' (1979, 152) account, is the direct offspring of the Primal Sun, *Yeba Haku*. From head to tail, the anaconda represents the siblings of *Yeba Haku* in order of birth, and hence in order of rank, with the head being the top rank and the tail being the lowest rank. These siblings are the ancestors of the *Bará*, *Barasana*, and the *Tatuyo*. The anaconda then entered the world from the east and traveled up the "Milk River (the Amazon?) to the Apaporís and thence to the Pirá-paraná [Rivers]—the middle of the world (Hugh-Jones 1979:152)." The anaconda

[12] Sometimes referred to as *Iuruparí* or *Juruparí*.

[13] *Socratea exorrhiza* (previously called *Iriartea exorrhiza*) is known as the paxiuba palm, or the walking palm, because it has unusual stilt roots that help to right the tree if it has fallen or been knocked over. When the trunk is horizontal, new roots will sprout from it to make it erect as old roots decay away. The result displaces the tree at a distance from its original germination location. The palm tree is also known to acquire numerous epiphytes (plants neutrally co-existing together, neither parasitic nor mutually beneficial) growing on it (Bodley and Benson 1980).

then turned its head to face the east and vomited up, or converted itself into, the tribes that are living along the rivers now. Whereas the *Kamayurá* myth might relate an ancient river-dwelling culture to the *jacaré* (crocodile), the *Barasana* myth more clearly relates a common ancestral river group to the anaconda.

The *Barasana* are not alone in their reverence for the anaconda. The *Ticuna* people,[14] who live near the southeastern border of Colombia and Brazil, continue to create bark-cloth paintings using vegetable dyes applied to the inner layer of the giant *Sapucaia* (*Sterculia striata*) tree (Guchte and Brewer 1992:8). In a conversation with Colombian anthropologist Margarita Chaves, Dr. Maarten van de Guchte (1992) learned that "traditional" bark paintings often feature a geometric layout of abstract designs, such as a cross with patterns resembling the skin of the anaconda.

Chaves explains that for the *Ticuna* and many other cultures along the Amazon River tributaries, the mythical anaconda serpent swam upstream, vomiting villages at specific locations, thereby providing the origin of life. Even today, many Indians say that the mythical anaconda resides in the beds of large rivers, serving as the source of water and sustaining life in those rivers (Guchte and Brewer 1992:11).

The *Ticuna* also have a myth about the origin of costumes made from bark cloth. In the myth recorded first by Curt Nimuendajú (1952:80), demons living in a cavern had eaten a group of traveling humans who all ate paca meat except for a pregnant woman, her husband, and their newborn child. Because they had not eaten the paca previously, they were spared from being eaten. Instead, they were instructed to climb a Ficus tree, stripping it of its bark in the process. Upon return of the young family to their tribe, they told their village what happened, and the tribe sought revenge by sealing the demons' cavern and smoking them out with burning peppers. The tribe did so, and after slaughtering the demons, they copied every one of their forms as costumes on the stripped bark (Guchte and Brewer 1992:20).

The significance of this myth is its direct mention of cave dwellers (demon or not) and the implied imitation of designs from those encountered in the mountains. The demon forms that were copied include the "phallic wind-demon, the female butterfly-demon, the catfish-faced water-demon, the tree-demon, the maize demon, and the monkey-demon" (Guchte and Brewer 1992:20). While it is unrealistic for "demons" to have looked like these descriptions of nature, it is theoretically plausible that the *Ticuna* copied cave drawings of these things left by a pre-existing culture.

Other bark-cloth paintings of the *Ticuna* are large enough to be hung as wall paintings. The most common types used in this context are bark-cloth paintings hung in the seclusion huts of young women before their puberty ceremony, and paintings hung on *maloca* (large circular multi-family hut) walls to delimit the space utilized for ritual dancing. A third use for the large bark-cloth paintings was to keep sacred wind instruments out of sight. Commonly reported themes found on these wall paintings include the face of the moon, a sun, a fish, a star, a deer, and

[14] Spelled *Tukuna* in this, and many earlier, references. The *Ticuna* speak a different language from the four languages surrounding Monte Alegre; their language is considered an isolate and is **not** related to the Tukanoan linguistic family.

geometric shapes from bands and triangles to rhomboids and circles: all of them placed in a quadrant divided by an X or cross (Guchte and Brewer 1992:22). Only the first four designs were reported to be traditional designs drawn in the old days, always at the top of the painting, depicting a new moon, a sun having a conventionalized human face, a star, and a deer as the symbol of vigilance (Nimuendajú 1952:76).

Of the four traditional designs of the Ticuna, three of them are themes associated with the sky, which returns us back to the Arawak-speaking groups. Another ritual of the Yawalapiti tribe is also shared by numerous tribes throughout Amazonia, and it is rooted in a peculiar superstition related to solar and lunar eclipses. In his documentary on the Yawalapiti ceremony, Matt Currington provides the following information:

> In the Yawalapiti community, the xamã (shaman) knows when there is a lunar eclipse because he always has a dream about it first. Then he warns the villagers what to do. They say the lunar eclipse is when the moon menstruates and can drip blood onto the earth. Also, the invisible spirits can infect the village if they are not honored. Honoring them includes a whipping ceremony by a descendant of a warrior who whips the young men during the time of the eclipse. The young men are believed to strip themselves of their old fears by being whipped and then running off into the pitch-black night. [Currington 2005]

Here we can see a clear cosmological connection between ancestral ceremonies, veneration of the spirits, and sky phenomena.

Culture and Cosmology Summary

The myths are not meant to be taken as historical accounts (nor can they be), but they allow us to sort through cultures that took over the land long after the period the rock art of Monte Alegre was made. In reviewing these cultures, we are searching for two things. The first is either direct or indirect references to pre-existing rock art, particularly red and yellow solar-themed paintings. The second is either direct or indirect references to a pre-ceramic highland culture that had endured or remained periodically mobile. Did the descendants of those Paleoindians survive with some hint of their ancestral heritage passing along or evolving into ceramic-age lowland cultures?

Arawakan cultures likely exerted the most widespread influence across Amazonia, as many groups appear to have adopted aspects of their social organization, subsistence strategies, community ceremonies, and religious idols. Archaeological evidence supports the theory that these widespread similarities stem from splinters of a common proto-culture, dispersed sporadically over a long period. In their rites-of-passage ceremonies, there are subtle references to stellar knowledge and superstitions, while sun icons blatantly adorn their chiefs. Arawakan groups also appear to have deeply incorporated the arts, including sacred musical instruments and gendered body paint. In their myths, however, we find metaphorical allusions to other groups.

From the non-Arawakan *Ticuna*, we learn of bark-cloth artistic traditions inspired by myths of cave dwellers. Few locations along the Amazon River contain caves, and while the *Ticuna* and their Arawak-speaking neighbors live closer to the Andes—much further west than the area of this research—their origin story of the anaconda sailing upriver to deposit villages along the Amazon points back toward the east. The most prominent mountainous terrain on the Atlantic side of the Amazon is in the *Guiana Highlands*, whose drainage system feeds into the *Orinoco River*, an area with a long history of *Carib* cultural influence.

From the *Carib* and other Orinoco and Guiana cultures, we trace an even stronger mythic connection between lowland ceramic-age cultures and highland cave, hill, and mountain dwellers. Most myths presented here, regardless of language or tribal affiliation, link mountains and their associated art to much older cultures. These were either directly ancestral or served as sources of artistic designs and inspiration. Trade connections are also evident, such as the *muiraquitã* greenstone amulets from the *Tapajós* culture in the Amazon reaching the *Orinoco* and even the *Island Caribs* in the Lesser Antilles. These amulets held great symbolic value, often associated with male–female relationships and the moon.

Although evidence is sparse, there are material and cultural indications—mythical, linguistic, and perhaps even genetic—suggesting links to ancient sun/moon-worshipping ancestors originating from painted mountain homelands. Many of these cultures connect ritual ceremonies, timed to ecologically significant seasons, to an ancestral mythical past.

How did they anticipate or determine when to hold rites-of-passage ceremonies, such as the female coming-of-age *pequi* festival or the male coming-of-age *Yurupary* ceremony? Were they held at fixed times each year based on the solar calendar, or were they variable, aligned with lunar phases? The next section delves deeper into how these traditions utilized naked-eye astronomy to determine the solar calendar. This practice underscores a rich ethnoastronomy tradition that offers shared cultural insights relevant to the rock art and archeoastronomy of *Monte Alegre*.

Ethnoastronomy

Theodor Koch-Grünberg (1905:58) was one of the first scholars to associate astronomy with mythology and oral narratives. However, as Edmundo Magaña (1986) points out, their concordance is not surprising because indigenous societies relied on oral tradition for the transmission of knowledge. And so, the real conundrum is the nature of their relationship. Yet, Magaña claims that scholars since Koch-Grünberg's time have grouped star myths with other symbolic narratives, and therefore sought universal meanings that inevitably led toward interpretations of human longings, passions, ethos, and deep-rooted cognitive symbols beyond any semblance of a practical or realistic representation.

Magaña had focused specifically on the economic basis of ethnoastronomy, since many cultures set their planting, harvesting, fishing, and hunting activities around star myths relating the appearance of one star or another with the heralding of, or direct cause of, a new season (Magaña 1986:403). Still other scholars have focused on the sociological roles that myths seem to convey as archetypical models for, or to discourage, certain social behavior (Goeje 1943; Kloos 1971; Levi-Strauss 1967; Roth 1915; Urton 1985) since cultures that envision constellations as animals tend to choose animals with consistent attributes highlighted by that culture.

An example of this ecological astronomy lore is provided by Alfred Métraux's account of the Chiriguano tribe, and how they utilize the stars as a *calendar* of cultural activities. "The appearance of the Pleiades, which the Chiriguano (Tupí) call 'a swarm of bees,' announces the harvest season. Their yearly course serves also to reckon time. The Chiriguano identify Scorpius with a fenced field, in the center of which is a miraculous spade of Aguara-tunpa. In the black skies near the Milky Way (the road of the rhea), they see a celestial ostrich (rhea), the head of which is the Southern Cross. The Magellanic Clouds are the ashes of a fire built by a couple who went to the sky (Métraux 1948a:483)."

By reviewing some ethnoastronomy interpretations for major sky observations, perhaps the fullest consideration currently possible for any astronomy legacy developed by the first Paleoindians can be obtained. This review, however, extends beyond the Amazon, to other Native people of the Americas, because we cannot be certain from where the Paleoindians of Monte Alegre came, nor their fate in the thousands of years Monte Alegre was abandoned.

The Sun

Fabian perhaps most accurately relates the mythic entities of the *Bororo* culture, near the border between Brazil and Bolivia, to the celestial entities in astronomy. Many *Bororo* myths feature *Meri*, the Sun, as clever, strong, and with the ability to resurrect his brother *Ari*, the Moon. Of this Fabian writes:

> In these attributes and through the kinship relationship between Sun and Moon, the Bororo are expressing metaphorically their conceptual understanding of the celestial sun and moon. That these cosmic brothers, who are also tricksters, are associated with heat is another, more overt link between the celestial and mythic entities. [Fabian 2001:31]

We also learn that the *Bororo* use the sun for village orientation and social movement, calendrically for day/time reckoning and for planning seasonal ceremonies and activities, and metaphorically for personal character and social interactions (Fabian 1992, 2001). In comparison, the *Ramkokamekra* people of the Eastern *Timbira* consider themselves dependent primarily on the sun, and to a lesser extent the moon, but they do not actively observe either one for calendrical purposes. Instead, they refer to the sun as "father" or "grandfather," and they petition to it for rain and to protect plants, animals, and newborn infants (Nimuendajú 1946:232).

Among the *Navajo* of southwestern North America, the moon determines monthly and nocturnal time, but "the sun's movement along the horizon is responsible for the seasons" (Griffin-Pierce 1992:74–75). The neighboring *Hopi* culture has sun priests and horizon "watchers" who utilize a solar observation calendar to determine planting and ceremonial cycles (Williamson 1984).

Moon

In *Patterns in the Sky* (2001: 37–39), Stephen Fabian recounts an experience with one of his *Bororo* informants. One evening, while observing the sky, he asked his *Paiwoe* clan elder informant about the moon. The elder replied that there was no moon that night (new moon) but that it would rise in the west the following evening. Fabian, perplexed by this statement—contrary to his understanding that both the sun and the moon rise in the east and set in the west—joined the elder the next evening. As the elder pointed to the west, Fabian saw a thin sliver of the waxing crescent moon low on the horizon just after sunset. The elder asserted that the moon was indeed rising in the west. Fabian observed that each day, looking to the west just after sunset, the moon appeared higher (and fuller) in the sky, as though it was rising from the west, even though it lowered into the western horizon shortly thereafter.

Fabian's example highlights how perspective, even on scientifically "certain" phenomena, is culturally interpreted. The *Bororo* perspective of the moon rising higher and fuller in the west serves as a cognitive tool for keeping better daily track of the moon's synodic period (its 29.5-day cycle of phases). This perspective may also assist in understanding how the moon causes solar eclipses. In contrast, modern societies typically explain the moon's phases using three-dimensional models of its orbit around the Earth and sun—a concept that traditional sky watchers often find abstract and difficult to relate to their direct observations.

For many cultures, lunar phases and positions are integral to agricultural and ritual practices. The *Bororo* use the lunar phases to determine when to harvest several important palm species for crafting cultural items or to plant specific domesticated crops. Additionally, the angle and fullness of the first visible waxing crescent moon serve as indicators of upcoming weather or human fate (Fabian 1992:92–96; 2001:46). Similarly, *Maya* farmers rely on lunar phases to decide the timing of butchering, harvesting, woodcutting, planting, and even sexual intercourse (Tedlock 1992:185).

Eclipses

Various Indian groups from southern Mexico through central South America still view eclipses with fear. They make as much noise as possible and encourage dogs to bark to frighten away the monster that they imagine is eating the sun or moon. Among some groups, they imagine a fight between the Sun and the Moon, and they encourage one or the other to fend off the aggressor (Kelley and Milone 2011:127).

The Bororo refer to a solar eclipse as the "death of Meri," the Sun, and that it is cause for "great terror" as an omen of death and misfortune (Fabian 2001:29). "The Timbira north of the Bororo envision a solar eclipse as the struggle between Sun and Moon, described in myth as two primordial males who discover each other and exist in a competitive relationship" (Fabian 2001:30). This second description is more approximate to the true cause of a solar eclipse that it strongly suggests the Timbira had discerned or inherited knowledge based on tracking the motions of the sun and moon at a time in the past when the myth was created. This should not be so easily dismissed because a solar eclipse can only occur during the period of an astronomical "new" moon, when the moon is otherwise not visible to us.

The *Timbira* also refer to a lunar eclipse as the "dying of the moon" (Fabian 2001:43).

Many cultures light fires, make commotions, sing and dance, shoot arrows into the sky to chase away the moon's assailant, and even offer human sacrifices to subside the presumed doom (Nimuendajú 1946:233). Unlike a solar eclipse, which can plausibly be determined by meticulous sky watchers of the moon, a lunar eclipse is caused by it passing behind the earth's shadow. This modern astronomy understanding requires knowledge of the earth as an orbiting body among the other two. This understanding goes beyond observation, requiring an entire conceptual paradigm shift. The animism of the *Timbira* accommodated by assuming the moon was dying. In similar association, the Maya believe a lunar eclipse signifies famine and sickness, so they also shout, shoot, and cause commotions to ward off the evil (Tedlock 1992:184).

Sun and Moon Halos

Regarding sun and moon halos specifically, the *Bororo* of Brazil's tropical savannah near Bolivia offer perhaps the most significant ethnographic analogy. In a simple entry of the *Enciclopédia Bororo I* (Albisetti and Venturelli 1962), "meri *ure marido mugudo tui-oku*" translates to "sun, circle/wheel around it," and "*ari ure marido mugudo tui-oku*" translates to "moon, circle/wheel around it" (Fabian 2001:80). The literal translations of these *Bororo* words for sun and moon halos specifically describe most of the Monte Alegre concentric circle rock paintings that I believe are representations of sun and moon halos.

Fabian further adds that most native peoples refer to such phenomena in relation to weather prediction (Fabian 2001:80–81). Quechua groups from Misminay predict the number of days before rainfall will occur based on the size of halos. Small ones precede rain by a few days; large ones precede rain by a day or so—of which their prediction accuracy has a scientifically confirmed account of 70% (Urton 1981:91,213 n.213).

Shooting Stars and Comets

The *Bororo* provide the most plausible ethnographic analogy for interpreting comets personified in the rock paintings at *Monte Alegre*. The *Bororo* refer to shooting stars as *aroe kodu*, meaning "flight of the soul," as they believe these celestial phenomena are omens linked to the theft of someone's soul by a malicious spirit. Comets, on the other hand, are described as "tailed stars" (Fabian 2001:82). The *Bororo* are not alone in envisioning shooting stars and comets as personified omens. The *Timbira* believe that a shooting star represents a star traveling to the other side of the sky to get married. They also attribute to comets the potential to set the world on fire (Nimuendajú 1946:233).

Despite these symbolic interpretations, there is no apparent cosmological association of comets with fire beyond their directional relationship with the sun. Comets are typically visible only when they approach the sun, millions of miles from Earth, and no heat is felt from them. They often appear white, blue, or greenish to the naked eye and are visible mostly when the sun is below the horizon. Conversely, meteors and bolides entering Earth's atmosphere create fiery streaks and sometimes audible rumbles or explosions, but these events are brief, lasting seconds to minutes.

Two other cultures enigmatically connect celestial events to catastrophic fires that destroyed much of the world in the distant past. The *Panoan*-speaking *Cashinawa* tribe, who live along the *Embira River* in Acre, Brazil, recounts a myth involving celestial destruction. According to the myth:

> "In the sky is a lake in whose depths is a hole which the stork keeps closed with his foot. When the stork goes fishing, rain falls. The *Cashinawa* attribute the destruction of the world and its first inhabitants to excessive rains, flooded rivers, and a general fire caused by a spark fallen from the sky. Some versions add that the sky crashed down and changed places with the earth. The first men were drowned; they either changed into animals or continued to live as spirits in the sky. The earth was repopulated by the progeny of the seven children of a sky woman who was struck by lightning. The babies were saved and brought up by a crab" (Métraux 1948b:684).

At face value, the myth seems nonsensical because a stork can't hold a lake in the sky. However, if the stork in the sky is a constellation that is no longer visible at night during the rainy season, it makes sense as a mnemonic for determining the beginning of the rainy season. Likewise, the notion of a crab raising children seems silly, but the *Cashinawa* themselves were known as the *Bat Indians*, so perhaps a different tribe were identified as the crab people.

The Ipuriná [Apurina] were another Arawakan group in Acre, Brazil. In the Ipuriná mythology, "a flood of boiling water was caused by the overflowing of the big pot of the sun, in which storks cooked all kinds of food … the moon is a young boy who grows alternately fat and lean; Orion is a beetle; the Pleiades, a serpent; the Southern Cross, a partridge; and the Hyades (including Aldebaran), a turtle (Métraux 1948b:684)."

These three stories about comets or sparks causing a catastrophic fire that ended much of the previous world are particularly peculiar, and perhaps they could relate to the Younger Dryas comet-impact hypothesis. Comets are usually visible for days to weeks. The two are easily distinguishable … until a comet hits the earth.

The Younger Dryas Comet-Impact Hypothesis

When I began writing this book, the comet-impact hypothesis was highly contested and deeply controversial. At the time, I did not anticipate having the time or space to address both sides of the debate thoroughly. However, in the many years it took to complete this book, evidence has continued to mount in support of the hypothesis (Powell 2022; Sweatman 2021). Although rock art or archaeoastronomy research alone may not provide sufficient evidence to resolve such a debate, I will briefly explain the hypothesis, as it might offer a potential correlation with the rock art of *Monte Alegre* in the future.

The *Younger Dryas* comet-impact hypothesis has been examined by numerous scholars worldwide. The hypothesis proposes that approximately 12,800 years ago, a fragment of a comet, believed to have originated from the *Taurid meteor shower* (Mahaney 2023; Usatov 2020), struck Earth. This impact is theorized to have triggered immense heat, leading to the rapid melting of a large portion of the *Laurentide* glacial ice sheet in North America (Mahaney and Somelar 2024). Such rapid melting is proposed to correspond with *Meltwater Pulse 1B*—a period of accelerated sea-level rise occurring between 11,500 and 11,200 years ago. This event coincides either with a delayed melting phase from the onset of the Younger Dryas or with its conclusion (Abdul et al. 2016; Bastos et al. 2022).

The hypothesized event suggests that the sudden release of massive quantities of meltwater from collapsing ice sheets into the oceans caused dramatic global sea-level changes (Firestone et al. 2007). These changes had far-reaching climatic and ecological impacts, which may have influenced the development of human cultures during this period. While the hypothesis remains the subject of ongoing debate, its potential relevance to interpreting the cultural and environmental context of the *Monte Alegre* rock art cannot be overlooked.

The comet is theorized to have exploded in a massive airburst, creating a 31-kilometer-wide impact crater beneath the Greenland ice sheet (Firestone et al. 2007) and causing widespread forest fires across parts of the continental United States at the time (Wolbach et al. 2018). Both claims remain contested, however (Holliday et al. 2023; Marlon et al. 2009; van Hoesel et al. 2012).

In addition, scholars have identified a sedimentary layer known as the *Younger Dryas Boundary* (YDB), which contains unusually high levels of platinum, iridium, nanodiamonds, shock-fractured quartz, melt glass, microspherules, and other quench-melted materials. These materials, found across multiple sites in the eastern United States (Bunch et al. 2012; Wittke et al. 2013), are known to form under extreme pressures and temperatures consistent with those produced by a cosmic airburst (Sweatman 2021).

The discovery of similar evidence at high latitudes in the Southern Hemisphere (Pino et al. 2019) supports the hypothesis that this was a widespread event (Israde-Alcántara et al. 2012; Kletetschka et al. 2018). Such an event likely triggered rapid cooling, contributed to megafaunal extinctions, and coincided with the disappearance of the *Clovis* culture in North America (Mahaney and Somelar 2024). It may even have impacted Paleoindian populations at *Monte Alegre* in the Amazon.

References

Abdul, N., R. Mortlock, J. Wright, and R. Fairbanks. 2016. Younger Dryas Sea level and meltwater pulse 1b recorded in Barbados reef crest coral Acropora Palmata. *Paleoceanography* 31 (2): 330–344.

Agostinho, P. 2009. *Mitos E Outras Narrativas Kamayurá*. 2nd ed. Salvador: Editora da Universidade Federal da Bahia (EDUFBA).

Albisetti, C., and A. Venturelli. 1962. *Bororo, Enciclopédia 1*. Mato Grosso do Sul: Museu Regional Dom Bosco, Campo Grande.

Bandeira, A. 2008. *Ocupações Humanas Pré-Históricas no Litoral Maranhense: Um Estudo Arqueológico sobre O Sambaqui do Bacanga Na Ilha De São Luís*. Maranhão: Departamento de Museu, Arqueologia, Etnologia, Universidade Federal do São Paulo.

Barata, F. 1950. *A Arte Oleira Dos Tapajó, I: ConsideraçÕEs Sobre a CerâMica E Dois Tipos De Vasos CaracteríSticos*. Belém: Instituto de Antropologia e Etnologia do Pará.

Barse, W. P. 1990. Preceramic occupations in the Orinoco River Valley. *Science* 250 (4986): 1388–1390.

Bastos, R. J. D. M. 2011. Leonardo, the flute: On the sexual life of sacred flutes among the Xinguano Indians. In *Burst of breath: Indigenous ritual wind instruments in lowland South America*, ed. J. D. Hill and J.-P. Chaumeil, 69–92. Lincoln: University of Nebraska Press.

Bastos, A. C., D. P. D'Agostini, A. E. Silva, P. S. Menandro, F. V. Vieira, G. C. Boni, V. S. Quaresma, and P. H. Cetto. 2022. Sedimentological and morphological evidences of meltwater pulse 1b in the southwestern Atlantic margin. *Marine Geology* 450:106850.

Bettendorf, J. F. 1909. Crônica Da Missão Dos Padres Missionários Da Companhia De Jesus Em O Estado Do Maranhão. *Revista do Instituto Histórico e Geografico* 72 (1): 5–682.

Boglar, L. 1958. Urn burial of the Brazilian Indians. *Acta Ethnographica* 6 (3–4): 347–355.

Bueno, L., and A. Isnardis. 2018. Peopling central Brazilian plateau at the onset of the Holocene: Building territorial histories. *Quaternary International* 473:144–160.

Bunch, T. E., R. E. Hermes, A. M. T. Moore, D. J. Kennett, J. C. Weaver, J. H. Wittke, P. S. DeCarli, J. L. Bischoff, G. C. Hillman, G. A. Howard, D. R. Kimbel, G. Kletetschka, C. P. Lipo, S. Sakai, Z. Revay, A. West, R. B. Firestone, and J. P. Kennett. 2012. Very high-temperature impact melt products as evidence for cosmic airbursts and impacts 12,900 years ago. *Proceedings of the National Academy of Sciences* 109 (28): E1903–E1912.

Campos, A. P. D. 1862. Breve Notícia Que Dá O Capitão Antônio Pires De Campos Do Gentio Que Há Na Derrota Da Viagem Das Minas Do Cuyabá E Seu Recôncavo. *Revista Trimestral do Instituto Histórico, Geográfico, e Etnográfico do Brasil* 5:437–449.

Cavalcante, P. B. 1988. *Frutas Comestíveis Da Amazônia*. 4nd ed. Companhia Souza Cruz Indústria e Comércio. Belém: Museu Paraense Emílio Goeldi.

Chaumeil, J.-P. 2007. Bones, flutes, and the dead: Memory and funerary treatments in Amazonia. In *Time and memory in indigenous Amazonia: Anthropological perspectives*, ed. C. Fausto and M. Heckenberger, 243–283. Gainesville, FL: University Press of Florida.

Civrieux, M. D. 1980. *Watunna: An Orinoco creation cycle*. Translated by D. M. Guss. San Francisco, CA: North Point Press.

Clement, C. R. 1999. 1492 and the loss of Amazonian crop genetic resources ii: Crop biogeography at contact. *Economic Botany* 53 (2): 203–216.

Clement, C. R. 2006. Fruit trees and the transition to food production in the Amazon. In *Time and complexity in historical ecology: Studies from the Neotropical lowlands*, ed. W. Baleé and C. L. Erickson, 165–185. New York: Columbia University Press.

Costa, F. 2009. *Arqueologia Das Campinaranas Do Baixo Rio Negro: Em Busca Dos Pré-Ceramistas Nos Areais Da Amazônia central*. Doctoral Dissertation. São Paulo: Departamento do Museu, Arqueologia, Etnologia, Universidade Federal do São Paulo.

Currington, M. 2005. Yawalapiti—Spirits of the eclipse. In *Tribal odyssey series*, 50. min, N. Geographic, general editor. National Geographic.

Daniel, F. J. J. 1841. Parte Segunda Do Thesouro Descoberto No Rio Amazonas- Noticia Geral Dos Indios Seus Naturaes, E De Algumas Nações Em Particular: Da Sua Fé, Vida, Costumes, E Das Cousas Mais Notaveis Da Sua Rusticidade. *Jornal do Instituto Historico Geographico Brasileiro* 3 (12): 422–441.

Davis, C., and A. G. Navarro. 2023. Plan of prehistoric Stilt Village in Maranhão Brazil may resemble the Pleiades. *Journal of Archaeological Science: Reports* 51:104123.

Denevan, W. M. 1966. *The aboriginal cultural geography of the Llanos De Mojos of Bolivia*. Berkeley, CA: University of California Press.

Denevan, W. M. 2001. *Cultivated landscapes of native Amazonia and the Andes*. Oxford: Oxford University Press.

Erickson, C. L. 2000. An artificial landscape-scale fishery in the Bolivian Amazon. *Nature* 408:190–193.

Erickson, C. L. 2008. Amazonia: The historical ecology of a domesticated landscape. In *Handbook of south American archaeology*, ed. H. Silverman and W. H. Isbell, 157–183. New York: Springer.

Fabian, S. M. 1992. *Space-time of the Bororo of Brazil*. Gainesville, FL: University Presses of Florida.

Fabian, S. M. 2001. *Patterns in the sky: An introduction to Ethnoastronomy*. Prospect Heights, IL: Waveland Press, Inc.

Ferraz, T., X. Suarez Villagran, K. Nägele, R. Radzevičiūtė, R. Barbosa Lemes, D. C. Salazar-García, V. Wesolowski, M. Lopes Alves, M. Bastos, and A. Rapp Py-Daniel. 2023. Genomic history of coastal societies from eastern South America. *Nature Ecology & Evolution* 7 (8): 1315–1330.

Firestone, R., A. West, Z. Revay, T. Belgya, A. Smith, and S. Que Hee. 2007. Evidence for a massive extraterrestrial airburst over North America 12.9 Ka ago. *AGU Spring Meeting Abstracts* 1:01.

Freire, J. R. 2001. Da "Fala Boa" Ao Português Na Amazônia Brasileira. *Amerindia* 8:1–83.

Gaspar, M. D., P. D. Blasis, S. Fish, and P. Fish. 2008. Sambaqui (Shell mound) societies of Coastal Brazil. In *Handbook of south American archaeology*, ed. H. Silverman and W. H. Isbell, 319–335. New York: Springer.

Gippert, J. 1993–2010. South American languages; historical distribution. In *Thesaurus Indogermanischer text- und Sprachmaterialien*, ed. Didactica. vol. 2013. TITUS, Frankfurt a/M.

Glück, H. 2000. *Metzler Lexikon Sprache*. 2nd revised ed. Stuttgart: Metzler.

Goeje, C. B. D. 1943. *Philosophy, initiation and myths of the Indians of Guiana and adjacent countries*, 44. Internationales Archiv für Ethnographie.

Góes, N. E., and J. B. Petersen. 2006. Political economy and pre-Columbian landscape transformations in Central Amazonia. In *Time and complexity in historical ecology: Studies from the Neotropical lowlands*, ed. W. Baleé and C. L. Erickson, 279–309. New York: Columbia University Press.

Gomes, D. 2002. *A Cerâmica Arqueológica Da Amazônia: Vasilhas Da Coleção Tapajônica Mae-Usp*. São Paulo: Oficial SP/Editora da Universidade de São Paulo.

Gonçalves, C. U. 2008. Os Piquizeiros Da Chapada Do Araripe. *Revista de Geografia* 25 (1): 88–103.

Griffin-Pierce, T. 1992. *Earth is my mother, sky is my father: Space, time, and astronomy in Navajo Sandpainting*. Albuquerque: University of New Mexico Press.

Grimes, B. F., R. S. Pittman, and J. E. Grimes, eds. 1996. *Ethnologue: Languages of the world*. Dallas, TX: Summer Institute of Linguistics.

Guapindaia, V. 2001. Encountering the ancestors: The maraca urns. In *Unknown Amazon*, ed. C. McEwan, C. Barreto, and E. Neves, 156–173. London: British Museum Press.

Guchte, M. V. D., and D. Brewer. 1992. *Masquerades and demons: Tukuna bark-cloth printing*. Urbana, IL: Krannert Art Museum—Board of Trustees of the University of Illinois.

Guedes, M. J. 1968. Brasil-Costa Norte. In *Cartografia Portuguesa vetustissima*. Ed. comemorativa do centenario da flotilha do Amazonas 1868-1968 ed. Rio de Janeiro: Ministerio da marinha, servico de documentacao geral de marinha.

Heckenberger, M. J. 2011. Deep history, cultural identities, and Ethnogenesis in the southern Amazon. In *Ethnicity in ancient Amazonia: Reconstructing past identities from archaeology, linguistics, and ethnohistory*, ed. A. Hornborg and J. D. Hill, 57–74. Boulder, CO: University Press of Colorado.

Heckenberger, M. J., and E. G. Neves. 2009. Amazonian archaeology. *Annual Review of Anthropology* 38:251–266.

Holliday, V. T., T. L. Daulton, P. J. Bartlein, M. B. Boslough, R. P. Breslawski, A. E. Fisher, I. A. Jorgeson, A. C. Scott, C. Koeberl, J. R. Marlon, J. Severinghaus, M. I. Petaev, and P. Claeys. 2023. Comprehensive refutation of the younger Dryas impact hypothesis (Ydih). *Earth-Science Reviews* 247:104502.

Hornborg, A. 2005. Ethnogenesis, regional integration, and ecology in prehistoric Amazonia: Toward a system perspective. *Current Anthropology* 46 (4): 589–620.

Hornborg, A., and L. Eriksen. 2011. An attempt to understand Panoan Ethnogenesis in relation to Long-term patterns and transformations of regional interaction in Western Amazonia. In *Ethnicity in ancient Amazonia: Reconstructing past identities from archaeology, linguistics, and Ethnohistory*, ed. A. Hornborg and J. D. Hill, 129–151. Boulder, CO: University Press of Colorado.

Hugh-Jones, S. 1979. *The palm and the Pleiades: Initiation and cosmology in Northwest Amazonia*. Cambridge, MA: Cambridge University Press.

Israde-Alcántara, I., J. L. Bischoff, G. Domínguez-Vázquez, H.-C. Li, P. S. DeCarli, T. E. Bunch, J. H. Wittke, J. C. Weaver, R. B. Firestone, and A. West. 2012. Evidence from Central Mexico supporting the younger Dryas extraterrestrial impact hypothesis. *Proceedings of the National Academy of Sciences* 109 (13): E738–E747.

Júnior, J. R. S., U. P. Albuquerque, and N. Peroni. 2013. Traditional knowledge and Management of Caryocar Coriaceum Wittm.(Pequi) in the Brazilian savanna, northeastern Brazil 1. *Economic Botany* 67 (3): 225–233.

Kelley, D., and E. Milone. 2011. *Exploring ancient skies: A survey of ancient and cultural astronomy*. 2nd ed. New York: Springer.

Kletetschka, G., D. Vondrák, J. Hruba, V. Prochazka, L. Nabelek, H. Svitavská-Svobodová, P. Bobek, Z. Horicka, J. Kadlec, and M. Takac. 2018. Cosmic-impact event in Lake sediments from Central Europe postdates the Laacher see eruption and marks onset of the younger Dryas. *The Journal of Geology* 126 (6): 561–575.

Kloos, P. 1971. *The Maroni River Caribs of Surinam*. Assen: Van Gorcum.

Koch-Grünberg, T. 1905. *Anfänge Der Kunst Im Urwald*. Berlin: Ernst Wasmuth.

Léry, J. D. 1992. *History of a voyage to the land of Brazil, otherwise called America*. Berkeley, CA: University of California Press.

Levi-Strauss, C. 1967. *Structural anthropology*. Translated by C. S. Jacobson and B. Grundfest 1. 2 vols. Garden City, NY: Doubleday/Anchor Books.

Levi-Strauss, C. 1969. *The raw and the cooked*. Translated by J. Weightman and D. Weightman. New York: Harper and Row.

Llamas, B., L. Fehren-Schmitz, G. Valverde, J. Soubrier, S. Mallick, N. Rohland, S. Nordenfelt, C. Valdiosera, S. M. Richards, and A. Rohrlach. 2016. Ancient mitochondrial DNA provides high-resolution time scale of the peopling of the Americas. *Science Advances* 2 (4): e1501385.

Magalhães, M. 1994. *Arqueologia De Carajás: A Presença Pré-Histórica Do Homen Na Amazônia*. Rio de Janeiro: Companhia do Vale do Rio Doce.

Magaña, E. 1986. South American ethno-astronomy. In *Myth and the imaginary in the New World*, ed. E. Magaña and P. Mason, 399–426. Latin American Studies. Amsterdam: CEDLA.

Mahaney, W. C. 2023. The Younger Dryas Boundary (Ydb): Terrestrial, cosmic, or both? *International Journal of Earth Sciences* 112 (3): 791–804.

Mahaney, W. C., and P. Somelar. 2024. The Encke comet impact/airburst and the Younger Dryas Boundary: Testing the impossible hypothesis (Ydih). *Geologos* 30 (1): 17–31.

Marlon, J. R., P. J. Bartlein, M. K. Walsh, S. P. Harrison, K. J. Brown, M. E. Edwards, P. E. Higuera, M. J. Power, R. S. Anderson, C. Briles, A. Brunelle, C. Carcaillet, M. Daniels, F. S. Hu, M. Lavoie, C. Long, T. Minckley, P. J. Richard, A. C. Scott, D. S. Shafer, W. Tinner, C. E. Umbanhowar Jr., and C. Whitlock. 2009. Wildfire responses to abrupt climate change in North America. *Proceedings of the National Academy of Sciences of the United States of America* 106 (8): 2519–2524.

Mason, J. A. 1950. The languages of South America. In *Handbook of south American Indians*, ed. J. H. Steward, vol. 6, 157–317. Bulletin. S. I. B. O.A. Ethnology, general editor. Washington, DC: Government Printing Office: Smithsonian Institution Bureau of American Ethnology.

McCallum, C. 2001. *Gender and sociality in Amazonia: How real people are made*. Oxford: Berg Publishers.

Meggers, B. J., and J. Danon. 1988. Identification and implications of a hiatus in the archeological sequence on Marajo Island, Brazil. *Journal of the Academy of Sciences* 78 (3): 245–253.

Meggers, B. J., and E. T. Miller. 2003. Hunters-gatherers in Amazonia during the Pleistocene-Holocene transition. In *Under the canopy: The archaeology of tropical rain forests*, ed. J. Mercader, 291–316. New Brunswick, NJ: Rutgers University Press.

Métraux, A. 1942. *The native tribes of eastern Bolivia and Western Matto Grosso*. Washington, DC: BAE Smithsonian Institution.

Métraux, A. 1948a. Tribes of the eastern slopes of the Bolivian Andes. In *Handbook of south American Indians: The tropical Forest tribes*, ed. J. H. Steward, vol. 3, 465–506. US Government Printing Office.

Métraux, A. 1948b. Tribes of the Jurua-Purus Basin. In *Handbook of south American Indians: The tropical Forest tribes*, ed. J. H. Steward, 657–799. Part 4. Tribes of the Western Amazon, US Government Printing Office.

Miller, E. T. 1999. A Limitção Ambiental Como Barreira À Transposição Do Period Formativo No Brasil. Tecnologia, Produção De Alimentos E Formação De Aldeias No Sudoeste Da Amazônia. In *Formativo Sudamericano: Homenaje a Albert rex González Y Betty J. Meggers*, ed. P. Ledergerber-Crespo, 331–339. ABYA-YALA Quito.

Moore, J. D. 2014. *A prehistory of South America*. Denver, CO: University Press of Colorado.

Morcote-Ríos, G., F. J. Aceituno, J. Iriarte, M. Robinson, and J. L. Chaparro-Cárdenas. 2021. Colonisation and early peopling of the Colombian Amazon during the late Pleistocene and the early Holocene: New evidence from La Serranía La Lindosa. *Quaternary International* 578:5–19.

Moreno-Mayar, J., L. Vinner, P. de Barros Damgaard, C. De La Fuente, J. Chan, J. Spence, M. Allentoft, T. Vimala, F. Racimo, and T. Pinotti. 2018. Early human dispersals within the Americas. *Science* 362:Eaav2621.

Navarrete, R. 1999. El Orinoco Medio. In *Arte PrehispáNico De Venezuela*, ed. M. Arroyo, L. Blanco, and E. Wagner, 34–51., G. a. d. A. Nacional, general editor. Caracas: Editorial Ex-Libris.

Navarro, A. G. 2022. Formoso Stilt Village: An incised-punctate/Arauquinoid tradition phase in Maranhão, Brazilian Amazon. *Advances in Anthropology* 12 (04): 149–168.

Neves, E. G. 2011. Archaeological cultures and past identities in the pre-colonial Central Amazon. In *Ethnicity in ancient Amazonia: Reconstructing past identities from archaeology, linguistics, and Ethnohistory*, ed. A. Hornborg and J. D. Hill, 31–56. Boulder, CO: University Press of Colorado.

Nimuendajú, C. 1939. *The Apinayé*. 1967 ed. Translated by R. H. Lowie. Anthropological Series. Oosterhout, NB: The Catholic University of America.

Nimuendajú, C. 1946. *The eastern Timbira*. Berkeley, CA: University of California Press.

Nimuendajú, C. 1949. *Os Tapajo 10*. Belem: Boletim do Museu Paraense Emilio Goeldi.

Nimuendajú, C. 1952. *The Tapajo*. Translated by J. H. Rowe. Berkeley, CA: Kroeber Anthropological Society Papers.

Noelli, F. S. 2008. The Tupí expansion. In *Handbook of south American archaeology*, ed. H. Silverman and W. Isbell, 659–670. New York: Springer.

Nordenskiöld, E. 1920. *The changes in the material culture of two Indian tribes under the Lnflumce of new surroundings*. Göteborg: Elanders.

Oberg, K. 1949. *The Terena and Caduveo of southern Mato Grosso*. Washington, DC: Smithsonian Institution.

Oliveira, R. C. D. 1968. *Urbanização E Tribalismo: A Integração Dos Índios Terêna Numa Sociedade De classes*. Rio de Janeiro: Zahar.

Oliver, J. R. 2008. The archaeology of agriculture in ancient Amazonia. In *Handbook of south American archaeology*, ed. H. Silverman and W. H. Isbell, 185–216. New York: Springer.

Palmatary, H. C. 1960. The archaeology of the lower Tapajós Valley, Brazil. *Transactions of the American Philosophical Society* 50 (3): 1–243.

Pané, R. 1999. *An account of the antiquities of the Indians*. Translated by S. C. Griswold. Duke University Press, Durham, NC.

Peters, C. 2000. Pre-Columbian Silviculture and indigenous Management of Neotropical Forests. In *Imperfect balance: Landscape transformations in the pre-Columbian Americas*, ed. D. Lentz, 203–223. New York: Columbia University Press.

Petersen, J. B., M. J. Heckenberger, and J. Wolford. 2001. Spin, twist, and twine: An ethnoarchaeological examination of group identity in native fiber industries from greater Amazonia. In *Fleeting identities: Perishable material culture in archaeological research*, ed. P. B. Drooker, 226–253. Occasional Paper, C. f. A. Investigations, general editor. Carbondale, IL: Southern Illinois University.

Pino, M., A. M. Abarzúa, G. Astorga, A. Martel-Cea, N. Cossio-Montecinos, R. X. Navarro, M. P. Lira, R. Labarca, M. A. LeCompte, V. Adedeji, C. R. Moore, T. E. Bunch, C. Mooney, W. S. Wolbach, A. West, and J. P. Kennett. 2019. Sedimentary record from Patagonia, southern Chile supports cosmic-impact triggering of biomass burning, climate change, and Megafaunal extinctions at 12.8 Ka. *Scientific Reports* 9 (1): 4413.

Potter, B. A., J. F. Baichtal, A. B. Beaudoin, L. Fehren-Schmitz, C. V. Haynes, V. T. Holliday, C. E. Holmes, J. W. Ives, R. L. Kelly, and B. Llamas. 2018. Current evidence allows multiple models for the peopling of the Americas. *Science Advances* 4 (8): eaat5473.

Powell, J. L. 2022. Premature rejection in science: The case of the younger Dryas impact hypothesis. *Science Progress* 105 (1): 368504211064272.

Prates, L., G. G. Politis, and S. I. Perez. 2020. Rapid radiation of humans in South America after the last glacial maximum: A radiocarbon-based study. *PLoS One* 15 (7): e0236023.

Prous, A., and E. Fogaça. 1999. Archaeology of Pleistocene-Holocene boundary in Brazil. *Quaternary International* 53:21–41.

Raghavan, M., M. Steinrücken, K. Harris, S. Schiffels, S. Rasmussen, M. DeGiorgio, A. Albrechtsen, C. Valdiosera, M. C. Ávila-Arcos, and A.-S. Malaspinas. 2015. Genomic evidence for the Pleistocene and recent population history of native Americans. *Science* 349 (6250): aab3884.

Reichel-Dolmatoff, G. 1971. *Amazonian cosmos: The sexual and religious symbolism of the Tukano Indians*. Chicago, IL: University of Chicago Press.

Reichel-Dolmatoff, G. 1978a. *Beyond the milky way: Hallucinatory imagery of the Tukano Indians*. Ucla Latin American Studies. Los Angeles, CA: UCLA Latin American Center Publications.
Reichel-Dolmatoff, G. 1978b. Desana animal categories, food restrictions and the concept of color energies. *Journal of Latin American Lore* 4 (2): 243–291.
Reichel-Dolmatoff, G. 1996. *The Forest within: The world-view of the Tukano Amazonian Indians*. Devon: Themis Books.
Ribeiro-dos-Santos, A. M., A. F. Vidal, T. Vinasco-Sandoval, J. Guerreiro, S. Santos, Â. Ribeiro-dos-Santos, and S. J. de Souza. 2020. Exome sequencing of native populations from the Amazon reveals patterns on the peopling of South America. *Frontiers in Genetics* 11.
Rodrigues, A. D. 1985. Evidence for Tupí-Carib relationships. In *South American Indian languages: Retrospect and Prospect*, ed. H. E. M. Klein and L. R. Stark, 371–404. Austin, TX: University of Texas Press.
Roosevelt, A. C. 1991. *Moundbuilders of the Amazon: Geophysical archaeology on Marajo Island, Brazil*. San Diego, CA: Academic Press.
Roosevelt, A. C. 1993. The rise and fall of the Amazon chiefdoms. *L'Homme* 255–283.
Roosevelt, A. C., M. Lima da Costa, C. L. Machado, M. Michab, N. Mercier, H. Valladas, J. Feathers, W. Barnett, M. Imazio, A. Henderson, J. Silva, B. Chernoff, D. S. Reese, J. A. Holman, N. Toth, and K. Schick. 1996. Paleoindian cave dwellers in the Amazon: The peopling of the Americas. *Science* 272 (5260): 373–384.
Rostain, S. 1994. The French Guiana coast: A key area in prehistory between the Orinoco and Amazon Rivers. Between St. Eustatius and the Guianas. *Publication of the St. Eustatius Historical Foundation* 3:53–97.
Rostain, S. 2008a. Agricultural earthworks on the French Guiana coast. In *Handbook of south American archaeology*, ed. H. Silverman and W. H. Isbell, 217–233. New York: Springer.
Rostain, S. 2008b. The archaeology of the Guianas: An overview. In *Handbook of south American archaeology*, ed. H. Silverman and W. H. Isbell, 279–302. New York: Springer.
Rostain, S., and A. H. Versteeg. 2004. The Arauquinoid tradition in the Guianas. In *Late ceramic societies in the eastern Caribbean*, Paris monographs in American archaeology, ed. A. Delpuech and C. L. Hofman, vol. 14., B. I. Series, general editor, 233–250. Oxford: Archaeo Press.
Roth, W. E. 1915. *An inquiry into the animism and folk-Lore of the Guiana Indians*, Thirteenth annual report of the Bureau of American Ethnology. Vol. 13. Washington, DC: Bureau of American Ethnology.
Schaan, D. 2001. Into the labyrinths of Marajoara pottery: Status and cultural identity in an Amazonian complex society. In *Unknown Amazon*, ed. C. McEwan, C. Barreto, and E. Neves, 108–133. London: British Museum Press.
Schaan, D. 2004. *The Camutins chiefdom: Rise and development of social complexity on Marajó Island, Brazilian Amazon*. Doctoral dissertation. Department of Anthropology, University of Pittsburgh, University of Pittsburgh.
Schmidt, M. 1914. *Die Paressi-Kabishi*. Leipzig: Baessler Archiv.
Seeger, A. 1981. *Nature and Society in Central Brazil: The Suya Indians of Mato Grosso*. Cambridge, MA: Harvard University Press.
Simões, M. F. 1969. The Castanheira site: New evidence on the antiquity and history of the Ananatuba phase (Marajó Island, Brazil). *American Antiquity* 34 (4): 402–410.
Skoglund, P., S. Mallick, M. C. Bortolini, N. Chennagiri, T. Hünemeier, M. L. Petzl-Erler, F. M. Salzano, N. Patterson, and D. Reich. 2015. Genetic evidence for two founding populations of the Americas. *Nature* 525 (7567): 104–108.
Smith, H. H. 1879. *Brazil the amazons and the coast*. London: Sampson Low, Marston, Searle, and Rivington.
Southey, R. 1810. *History of Brazil: Part the first*. London: Longman, Hurst, Rees, and Deme, Paternoster Row.
Steward, J. H., and L. Faron. 1959. *Native peoples of South America*. New York: McGraw-Hill.
Sweatman, M. B. 2021. The younger Dryas impact hypothesis: Review of the impact evidence. *Earth-Science Reviews* 218:103677.

Tedlock, B. 1992. *Time and the Highland Maya*. Revised ed. Albuquerque: University of New Mexico Press.

Tyacke, S. 1980. English charting of the river Amazon C. 1595-C. 1630. *Imago Mundi* 32:73–89.

Urton, G. 1981. *At the crossroads of earth and sky: An Andean cosmology*. Austin, TX: University of Texas Press.

Urton, G. 1985. *Animal myths and metaphors in South America*. Salt Lake City, UT: University of Utah Press.

Usatov, M. 2020. Main Belt asteroid as a possible younger Dryas impactor. *Astronomische Nachrichten* 341 (8): 734–740.

van Hoesel, A., W. Z. Hoek, F. Braadbaart, J. van der Plicht, G. M. Pennock, and M. R. Drury. 2012. Nanodiamonds and wildfire evidence in the Usselo horizon postdate the Allerød-younger Dryas boundary. *Proceedings of the National Academy of Sciences* 109 (20): 7648–7653.

Villas Boas, O., and C. Villas Boas. 1970. *Xingu: The Indians, their myths*. Translated by S. H. Rudge. New York: Institute of Social Anthropology, Oxford University. Farrar, Straus and Giroux, Inc.

Wallace, A. R. 1889. *A narrative of travels: Amazon and Rio Negro, with an account of the native tribes*. 2nd ed. London: Ward, Lock and Co.

Waters, M. R. 2019. Late Pleistocene exploration and settlement of the Americas by modern humans. *Science* 365 (6449): eaat5447.

Wilbert, J., and K. Simoneau, eds. 1978. *Folk literature of the Ge Indians*. Vol. 44. Los Angeles, CA: UCLA Latin American Center Publications.

Wilbert, J., and K. Simoneau, eds. 1987. *Folk literature of the Gê Indians*, Ucla Latin American studies. Vol. 44. Los Angeles, CA: UCLA Latin American Center Publications.

Williamson, R. A. 1984. *Living in the sky: The cosmos of the American Indian*. Boston, MA: Houghton Mifflin.

Wittke, J. H., J. C. Weaver, T. E. Bunch, J. P. Kennett, D. J. Kennett, A. M. T. Moore, G. C. Hillman, K. B. Tankersley, A. C. Goodyear, C. R. Moore, I. R. Daniel, J. H. Ray, N. H. Lopinot, D. Ferraro, I. Israde-Alcántara, J. L. Bischoff, P. S. DeCarli, R. E. Hermes, J. B. Kloosterman, Z. Revay, G. A. Howard, D. R. Kimbel, G. Kletetschka, L. Nabelek, C. P. Lipo, S. Sakai, A. West, and R. B. Firestone. 2013. Evidence for deposition of 10 million Tonnes of impact spherules across four continents 12,800 Y ago. *Proceedings of the National Academy of Sciences* 110 (23): E2088–E2097.

Wolbach, W. S., J. P. Ballard, P. A. Mayewski, V. Adedeji, T. E. Bunch, R. B. Firestone, T. A. French, G. A. Howard, I. Israde-Alcántara, and J. R. Johnson. 2018. Extraordinary biomass-burning episode and impact winter triggered by the younger Dryas cosmic impact∼ 12,800 years ago. 1. Ice cores and glaciers. *The Journal of Geology* 126 (2): 165–184.

Zucchi, A. 1985. Evidencias Arqueológicas Sobre Posibles Grupos De Lenguas Caribe. *Antropológica* 63–64:23–44.

Conclusion

Long before Galileo peered through a telescope and challenged religious doctrine on the nature of the cosmos, Amazonian rock artists may have painted notations of celestial cycles as signs anticipating the recurrence of seasonal resources. These early practices offer a glimpse into pre-scientific predictions rooted in objective reality. This book presents research on the first people to inhabit a group of hills along the banks of the lower Amazon River over 13,000 years ago. Archaeological excavations have unearthed sandstone lithic tools, natural earth pigments, and quarry-like features dating to this initial cultural period of habitation. These findings suggest that this pioneering culture created rock art pictographs with potential ecological or astronomical significance and may have engaged in landscape modifications.

The radiocarbon dates for this period of cultural activity—13,236–12,755 calibrated years before present (cal yr BP)—align with the *Clovis* period in North America and the latter half of the *Bølling-Allerød* warm period. This era concludes at the onset of the *Younger Dryas* cool period at the end of the last Ice Age, marking the Pleistocene/Holocene boundary. The research proposed in this book indicates that the depictions, positions, and locations of pictographs suggest the artist(s) recorded sky observations of the sun and star groups, particularly along the horizon. These observations may have been used to track ecological transitions throughout the year. If true, this practice would have allowed these early inhabitants to acclimate to their new environment in the hills, enabling them to predict the Amazon River's seasonal flooding and the resulting changes in aquatic and terrestrial plants and animals.

The adaptive cognition inherent in drawing stellar constellations as zoomorphic or mnemonic representations of transient ecological resources could have been a universal practice, applicable to any environment on Earth. To ancient peoples, this might have been as intuitive as digging a hole to find water. The key was not merely recognizing shapes in the constellations but connecting those constellations meaningfully to the resources vital to survival in their specific landscape. The

C. S. Davis, *Before Sunset*, Conflict, Environment, and Social Complexity,
https://doi.org/10.1007/978-3-031-93373-8

constellation's prominence at a particular time in the sky marked a corresponding period in the ecological cycle.

In Amazonian cultures, celestial delineations often go beyond the stars, incorporating nebulae and dark spaces. A star group appearing on the eastern horizon at sunset will also pass through the zenith at midnight, dividing the annual cycle of stars into predictable segments. This cognitive framework is the basis of the zodiac constellations known today, with their 12 (or 13) divisions. These imagined sky figures endure by being mythologized into stories and fables.

Occasionally, individual planets serve as variable agents—birds, arrows, or people—who modify the celestial narrative. Much like a Disney fairytale, astronomical myths blend animate and inanimate characters to advance their stories. This dynamic approach allows cultures to embed festivals and rites of passage into their sky lore, ensuring that traditions remain adaptive while retaining historical stability.

Reviewing the data and major theories presented in this book paints the following inferences and suppositions:

- DATA: Earliest radiocarbon dates from excavations at Monte Alegre calibrate to a range from the Bolling-Alerod warming period to the beginning of the Younger Dryas
  - THEORY: Perhaps more amiable climate encouraged ancient pioneers to venture into unexplored lands, and upon traveling down the Amazon River, they came upon the hills of Monte Alegre rising above the inundated lands.
  - The prospect of a dry landing place where lithic tools could possibly be replenished would have been a strong draw for Paleoindians of the stone age.
- DATA: The earliest inhabitants painted numerous images of dots, circles, rayed circles, concentric circles, rayed concentric circles, and concentric dots, many which face horizon sunsets or sunrises throughout the hills. This category (motif) of pictographs therefore only appears to have been drawn within the path of the sun within a solar year, from northern (summer) solstice to southern (winter) solstice.
  - THEORY: This motif category correlates strongly with the sun. There are practically no images that unambiguously depict phases of the moon. If this motif was indeed meant to represent the sun, the horizon bounds (cardinal directions) by which these drawings are limited are therefore functional (or more specifically, notational) not simply creative drawing or doodles on a whim, which we would expect to find anywhere, in any orientation—or at least in random orientations. So, notation is likely; the knowledge it provides is that the sun's horizon position determines seasonal changes in the environment, and the ecological cycles that Paleoindians relied on for their survival. The sun motif possibly helped these ancient artists to gage the progression of time.
- DATA: Many of the caves—especially in *Serra da Paituna* (*Abrigo do Coruja, Gruta 15 de Março*, possibly even *Pedra Pintada*)—depict a prominent

zoomorphic pictograph in comparison to other pictographs in the same cave. Material evidence or direct sightings of animals still inhabiting those caves today are usually the same associations referenced in the name of the cave.

  - THEORY: Upon exploration of the land, its caves, and rockshelters, Paleoindians began to associate certain areas as the "homes" of certain animals. A "patron" animal image was drawn in the cave as an identifier to "label" the cave, perhaps in the process of "landscape learning."

- DATA: Although this segment of the Amazon River is largely considered to just have a long, wet season and a short, dry season, different groups of animals come into and out of the region in cyclical egresses.
  - THEORY: The earliest inhabitants would find these patterns to be vitally important to learn because once they can expect certain animals to be available at certain times of the year, they can schedule their activities for optimal foraging effort. To remember these patterns, the rock art could have served as record, or an aide memoire for teaching by memory. And, perhaps, these animals may have been planted within the solar motif context, serving as a calendar or almanac.
- DATA: The rock art that spans the distance from *Serra da Lua* to *Serra do Sol* faces sunsets spanning an entire year.
  - THEORY: Allowing for (at least) some pictographs to annotate the animal egresses that cycle through the year.
- INFERENCE: The repeated recording of careful sky observations using rock art may have led to successful adaptation of Paleoindians to the landscape through a gained predictive modeling of their environment. They may have known this ancient "scientific" strategy before arriving, but here we find the earliest physical evidence that it was implemented in South America, just before the Younger Dryas.

The manufacturer's authorised representative in the EU is Springer Nature Customer Service Centre GmbH, Europaplatz 3, 69115 Heidelberg, Germany. If you have any concerns regarding our products, please contact ProductSafety@springernature.com

Printed and bound by CPI Group (UK) Ltd, Croydon, CR0 4YY
30/09/2026
02225501-0005